Small Fruits in the Home Garden

By
R.M. Skirvin
A.G. Otterbacher
J.D. Kindhart
M.M. Kushad
K.D. McPheeters
S.M. Ries
R.A. Weinzierl

Circular 1343
(1997 Completely Revised)
Department of Natural Resources and Environmental Sciences
and Department of Crop Sciences
University of Illinois at Urbana-Champaign

UNIVERSITY OF ILLINOIS EXTENSION

College of Agricultural, Consumer and
Environmental Sciences

Issued in furtherance of Cooperative Extension Work, Acts of May 8 and June 30, 1914, in cooperation with the U.S. Department of Agriculture. Dennis Campion, Interim Director, Cooperative Extension Service, University of Illinois at Urbana-Champaign. The Cooperative Extension Service provides equal opportunities in programs and employment.

ISBN 1-883097-20-7

6M—2-98—UI Printing Division—CF/PC

Acknowledgements:

Fig. 3: Dr. Oscar Schubert; photo on page 20, Fig. 16, Fig. 17, and photo on page 34: USDA; Fig. 21, Fig. 24, Fig. 25, Fig. 26, Fig. 30; Rutgers University. The rest are provided by the authors or previous authors.

This publication was last published as University of Illinois Circular 935. It was prepared through various revisions by H.C. Barrett, J.W. Courter, C.C. Doll, D.B. Meador, A.G. Otterbacher, D. Powell, M.C. Shurtleff, R.K. Simons, R.M. Skirvin, and C.C. Zych. The section on landscaping with small fruits was co-authored by S. Morrissey.

Table of Contents

For further information...

For assistance in selecting the proper species and cultivar, order catalogs from reputable nurseries and read the plant descriptions carefully. University of Illinois College of Agricultural, Consumer and Environmental Sciences has a variety of resources available on growing fruits and related practices. Contact the U of I Extension for applicable circulars and fact sheets, or contact the nearest local Extension office. Inquiries can also be made to the College of ACES's Information Technology and Communication Services office, 1301 W. Gregory Drive, Urbana, Illinois 61801, (217)333-4780.

Planning the Small Fruits Garden

Growing small fruits in the home garden offers many advantages to a family that is willing to provide the space and care plants require. All major small fruits—including strawberries, raspberries, blackberries, blueberries, currants, gooseberries, and grapes—can be grown successfully in most parts of Illinois. These plants tolerate small spaces and are popular even for small city lots. Space limitations can be overcome by fitting small fruit plants directly into the overall landscape in shrub borders, screen plantings, arbors, hedges, patios, or perennial gardens.

A well-planted garden supplies fresh fruit from early spring to the first killing frost in the fall. The fruits produced are enjoyed for their pleasant taste as well as for their dietary value as sources of vitamins, minerals, acids, and anti-cancer agents. Fruit of the best cultivars, harvested at the peak of their season, cannot be matched in the market regardless of price. Surplus production can be sold as an income supplement or canned, frozen, or dried for use year-round.

Aside from the benefits of superior quality, the care and cultivation of small fruits at home can provide much pleasure and satisfaction. Careful selection of early- and late-season cultivars of different small fruits will supply fresh fruit over the longest possible season. Success, of course, also depends on careful attention to cultural details. This circular describes cultural techniques and cultivars for reliable production in Illinois home gardens.

(left) This hedge of thornless blackberries illustrates how a small fruits garden can be incorporated directly into an overall landscape.

Size of Planting

The number of plants the home gardener can grow is determined by the space and time available as well as the needs of the family. Home fruit plantings should be limited in size, especially if the primary objective is only to supply fresh fruit for home use. Large plantings may be justified for local sales or a 4-H project. The best advice is to plant no more than you can properly care for. Neglected plants produce low yields of poor-quality fruit, harbor destructive insects and diseases, and can be unsightly.

Plant spacings, approximate yields, and a suggested number of plants for a family of five are given in Table 1 (inside front cover).

Although there is no sharp demarcation between northern, central, and southern Illinois, some cultivars definitely perform better at a given latitude. For the purposes of this book, we have defined the regions as N = region north of Interstate 80; C = central region between Interstate 80 at the north and Interstate 70 at the south; and S = region south of Interstate 70.

The yields shown are dependent upon proper cultivar selection and good management. The size and layout of the garden may vary according to the selection of fruits desired and the space and location available. When arranging small fruits in the garden, place the taller-growing fruits such as trellised grapes north of lower-growing fruits such as strawberries to prevent shading.

Location

It is a good idea to locate the small fruit garden near the house. If large trees are nearby, locate the garden to the south of them, if possible. There is an advantage in planting adjacent to the vegetable garden for convenience of doing cultural chores. However, sometimes the most convenient location does not have the most desirable exposure, soil, or water drainage.

The site should have reasonably fertile soil and be well drained. Avoid areas that collect water after a rain. A moderately elevated or sloping site will reduce losses from late spring frosts and provide better water drainage. Exposure to full sunlight is preferred, although most small fruits will grow in partial shade. Black raspberries and gooseberries can stand more shade than other small fruits. Where possible, take advantage of natural windbreaks such as buildings and hedgerows to protect the planting from severe prevailing northwest winter winds.

Strawberries and brambles should not be planted on a site where tomatoes, potatoes, peppers, eggplant, melons, okra, peas, beets, or roses have been grown in the past three to five years. Verticillium wilt is a disease common to all these crops, and if they are not rotated with nonsusceptible crops, severe losses can result. There are Verticillium-resistant strawberry and raspberry cultivars available; some of these cultivars are listed in the appropriate section later in this circular.

Preparing the Soil

Most small fruit plants will occupy the same location at least three years and as long as thirty or more years. Therefore, it is desirable to build up the soil fertility of the proposed planting site before planting. Planning one or two years ahead can also help to reduce weed problems.

All small fruit plants benefit from the addition of organic matter to the soil. If well-rotted manure that is free of weed seed is available, incorporate four bushels (or 100 pounds) per 100 square feet (1,000 to 1,500 pounds per 1,000 square feet) in the summer or fall before planting. Similar levels of compost, decomposed leaves, or lawn clippings can also increase the soil's organic-matter level. Use only lawn clippings that have not been treated with herbicides. It is a good idea to add ten to fifteen pounds of 10-10-10 fertilizer per 1,000 square feet when leaves are used. Thoroughly work the organic material into the soil. In September, sow a cover crop such as annual rye at three pounds of seeds per 1,000 square feet to protect the soil during winter. Turn the crop under in spring to improve the soil.

To reduce weeds, plant and cultivate row crops for one or two years prior to planting small fruits. Avoid the Verticillium-susceptible crops listed earlier. Also, avoid sites where herbicides with long carry-over periods have been used recently. Regular cultivation and hoeing will be necessary to control weeds. The cultivation helps improve soil conditions by mixing organic matter in the soil and increasing aeration.

If sod must be turned under, it should be done in fall to allow decomposition to begin. Sod often harbors grubs (insect larvae) that feed on roots. Most of the grubs die if the sod is turned under and the ground cultivated at least one year prior to planting.

All small fruits, except blueberries, grow satisfactorily in a soil pH range of 5.5 to 7.5. Blueberries require a pH of 4.8 to 5.2 for best growth. The pH measurement refers to the relative acidity or alkalinity of the soil: 7 is neutral, 6 to 7 is slightly acid, and 7 to 8 is slightly basic (alkaline). Horticulture Fact Sheet FL-13-81, "Taking Soil Samples in the Yard and Garden," provides instructions on how properly to obtain soil samples. Soil-testing services are often listed in the telephone book Yellow Pages under "soil testing." Contact the nearest regional Extension office for help in interpreting the soil-test results, if help is needed.

Prior to planting, work the soil as thoroughly as you would in planting a vegetable garden. The soil should be well pulverized and moist, and soil granulation should be no larger than a pea.

Planting Stock

Healthy, vigorous plants are essential for establishing a successful small fruit planting. It is wise and, in the long run, cheaper to buy the best plants available. The disadvantages of poor planting stock can never be overcome. Reputable nurseries supply disease-free and true-to-name plants. The state certificate of nursery inspection is your assurance of healthy plants. Strawberry, raspberry, and blackberry plants should be "virus-free" at the time of purchase. Many of these plants are now supplied from tissue-culture propagation. Tissue-culture plants may cost more than traditionally propagated plants but are worth the extra cost.

Obtain catalogs from several nurseries. While some nurseries specialize in one or two types of small fruits (specialty nurseries), others are more general and supply most small fruits. Orders should be placed early to obtain desired cultivars. November or December is not too early to order plants for the following spring. The delivery date and method of shipment should be specified when placing the order.

One-year-old plants of medium to large size are generally the best to buy. The added cost for older or extra-large plants is seldom justified. Blueberry plants are an exception. It is best to purchase two-year-old container-grown plants. Two-year-old transplants are also good.

Choosing Cultivars

Cultivated varieties (cultivars) for home small-fruit plantings should be selected for their high quality for eating fresh, preserving, or both. Many cultivars of high-quality small fruits are not suited to commercial production, so the only source of these quality fruits may be your own garden. Disease resistance and winterhardiness also are very important considerations. Careful selection of early- and late-maturing cultivars will provide fresh fruit over a long harvest season. The use of several cultivars helps to ensure a successful planting; one cultivar may perform well in one location and poorly in others. The cultivars suggested in this circular are generally adapted to Illinois conditions. Special notation is made when a particular cultivar is best suited to a particular region. In addition to the suggested cultivars, gardeners can purchase and compare one or two new cultivars.

Care of Plants on Arrival

Most plants are dug by nurseries in late fall or early spring while the plants are dormant, stored under refrigeration, and then shipped dormant during the correct season. Such plants, when handled properly, are as good as freshly dug plants.

Packages should be opened and the plants examined as soon as they arrive. If they smell sour and rotten, they should be replaced. Healthy plants have an earthy smell; roots should be supple and white. It is a good idea to peel back the wrapping to let excess heat escape from the package. *Do not let your plants*

1 Heeling-in strawberry plants. The bundles are opened, the plants spread out one deep in the trench, and moist earth thrown back and packed around the roots. Here they will keep fresh several days until conditions are favorable for planting.

dry out. If the plants are dry when they arrive, soak the roots in water for one or two hours and plant immediately, if possible. If planting is delayed more than one day, place the plants in cold storage or in a refrigerator (32° to 40° F) that is not being used to store fruits and vegetables, or "heel- in" the plants (Fig. 1).

For cold storage, moisten the roots if they are dry, but be careful not to get them too wet, or the plants may mold and rot. Plants in plastic bags may be kept satisfactorily for a week in your home refrigerator. Avoid storing plants in refrigerators that also contain fresh fruits (especially apples and pears) because as the fruit ripens, it produces a gas (ethylene) that can injure or kill plants. Do not allow plants to freeze.

To "heel-in" plants, select a location that is well drained, shaded, and protected from the wind. Dig a trench deep enough to permit covering the roots and long enough to place all of the plants side by side, one plant deep. After positioning plants in the trench, firm the soil over the roots. Do not cover the crowns of strawberry plants. Water the plants thoroughly and keep them shaded until they are ready to plant. Do not leave the plants heeled-in any longer than absolutely necessary. Their roots will begin to grow, and transplanting them will destroy some of their new feeder roots.

Irrigation

Lack of rain while new plants are becoming established—during bloom and harvest and during late summer and fall when fruit buds are forming—can reduce the quantity and quality of fruit. For optimal growth, most small fruits require at least one inch of water per week during the growing season. Irrigation to supplement rainfall to this level is especially important for soils subject to drought, such as sandy soils, or soils with a shallow hardpan that restrict the development of a deep-root system.

If possible, locate the small fruit garden where water is readily available for irrigation. Sprinklers, porous soaking hoses, and drip or trickle hoses are suitable for applying water. Irrigate to thoroughly wet the soil to the depth of the roots. Shallow watering encourages shallow root growth and is of little value (and may even be harmful) to long-term plant survival.

Pruning Tools

Correct pruning and training are necessary for good production of the brambles, blueberries, currants, gooseberries, and grapes. The necessary tools are not complex or expensive (Fig. 2). The hand clipper is used to cut back small branches and lateral shoots. There are two types of hand shears: the by-pass type cuts cleanly with a scissorslike action; the anvil type cuts with a crushing action. Most horticulturists agree that by-pass shears are superior to the anvil type. The lopping or long-handled shears is needed for larger branches and canes that cannot be cut with the hand shears. The bramble hook is a specialized tool for removing entire canes from the brambles. A pruning saw may be needed for grapes and blueberries.

Pruning tools work best when they are sharp and clean. Tools should be cleaned after each use and their cutting edges should be wiped with an oily cloth to prevent rust. The cutting edges should be kept sharp to make smooth cuts that will heal quickly.

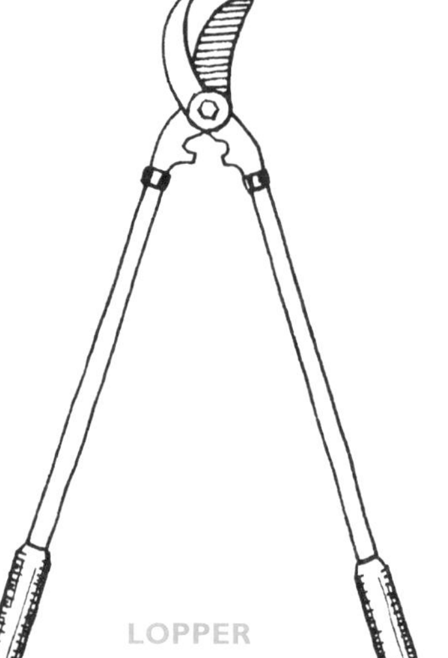

2 LOPPER

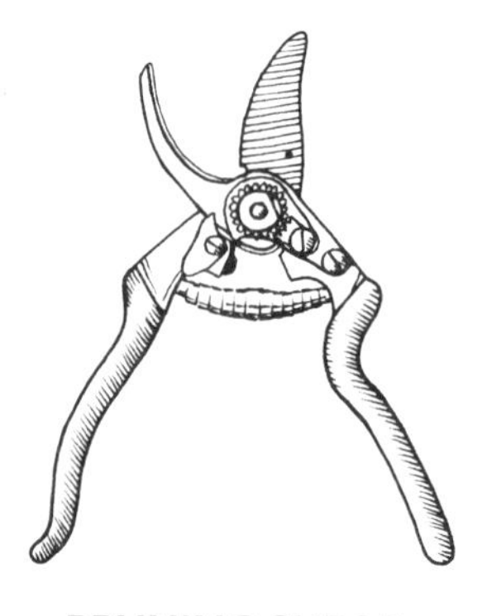

PRUNING SHEAR

The Home Landscape

Individual small fruit plants or groups of small fruit plants can be included in the landscape to provide fresh fruit when space is not available for a defined fruit garden (Fig. 3). Strawberries—particularly the disease-resistant, day-neutral types that bear fruit throughout the summer—can be useful for edging or ground-cover plantings. Grape arbors or hedge plantings of erect blackberries, raspberries, or blueberries can be used effectively to partially screen or separate parts of the lawn or garden. Blueberries are especially well adapted to the home landscape: they have lovely white bell-shaped flowers in the spring, blueberries in the summer, and attractive red foliage in the fall. With a little imagination and careful planning, small fruit plants can be combined with other plants to result in an eye-pleasing, appetizing landscape planting. This topic is discussed in more detail beginning on page 65.

3 Red currant growing in a backyard fence.

Strawberries

Strawberries are the most popular of the small fruits. They are the first to ripen in the spring and are highly nutritious. A single cup of strawberries supplies more than the minimum daily requirement of Vitamin C and is a tasty way to fight cancer. Choosing disease-resistant cultivars usually results in the production of satisfactory crops in a home garden with minimal use of pesticides.

Three types of strawberries are grown in Illinois: spring or June bearing, everbearing, and day neutral. June bearers produce their crop in a two- to three-week period in the spring. The everbearing strawberries usually produce three flushes of flowers and fruits throughout the growing season—spring, summer, and fall. Day neutrals will flower and fruit continually through the growing season. Many cultivars of strawberries are suitable for culture in Illinois.The cultivars listed in Table 2 (page 19) have proven their adaptability to Illinois. New cultivars are introduced regularly, so the list is subject to change.

Soil

Strawberries will grow satisfactorily in most garden soils, but they require a relatively high level of soil fertility for optimal production. The soil pH should be between 5.5 and 6.5. Livestock manure, preferably well-rotted, may be applied the year before planting (see page 16). If manure is not available, compost and commercial fertilizer can be added when preparing the soil. Apply fifteen to twenty pounds of 10-20-20 fertilizer, or equivalent, per 1,000 square feet, and work the fertilizer into the top six to eight inches of soil.

4 **Proper planting depth.** (right) The center plant is set correctly, with the soil just covering the tops of the roots. The plant on the left is set too shallow; the plant on the right, too deep.

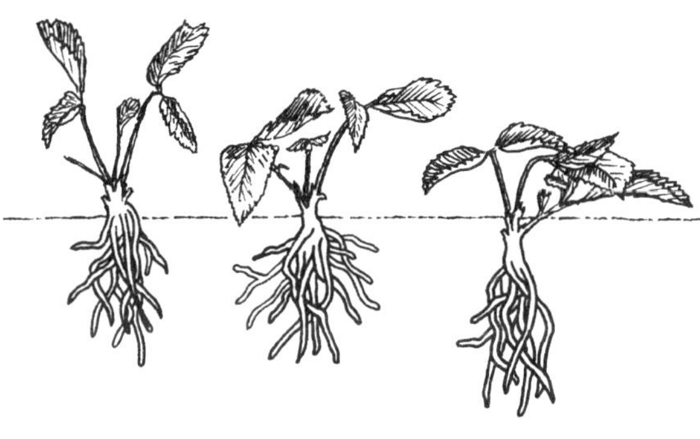

5 (left) Young plants soon after planting, with runners.

6 Runners can be made to grow in **"matted rows"** by training most of the runners into the row and rototilling out the others (see photo right and diagram below).

7 **Spaced row plantings** (below) These spaced plantings were made by limiting the number of runners to two or four, or all runners can be removed to yield hills (Fig. 8).

matted row

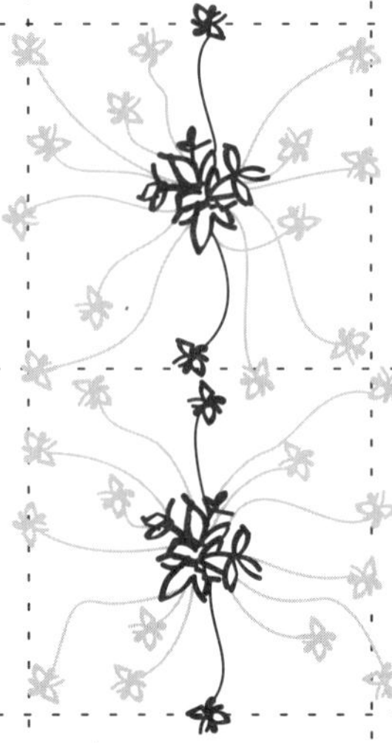

spaced row: two runners

spaced row: four runners

Planting and Spacing

Strawberries should be planted as soon as the ground can be prepared in the spring. Do not plant if the soil is wet. Planting is best done in March or April in Illinois to allow the plants time to become well established before the hot summer weather begins. If possible, the plants should be set during cloudy weather or during the late afternoon or evening. Set the plants to the proper depth and apply one pint of water per plant (Fig. 4). Within four to five weeks, mother plants will produce runners and new daughter plants (Fig. 5).

The matted-row system (Fig. 6) is the most popular method for growing June-bearing (standard) cultivars in Illinois. The plants should be set eighteen to thirty inches apart in rows three to four feet apart. The daughter plants are allowed to root freely to become a matted row no wider than two feet.

8 Hill plantings for day neutrals.

Spaced-row systems (Fig. 7) limit the number of daughter plants that develop from a mother plant. Under this system, the original mother plants are spaced the same as before, but the daughter plants are spaced to root no closer than four inches apart. All other runners are pulled or cut from the mother plants. Such spacing gives optimal growing conditions because strawberry rows can often be too dense for good production. Spaced-row culture requires more care than matted-row plantings, but higher yields, larger berries, and fewer disease problems may justify the extra effort.

The hill system (Fig. 8) is the best method to grow everbearing and day-neutral cultivars. All runners are removed so that only the original mother plant is left to grow. Runners develop from the same region as flower stalks, so runner removal enables the mother plant to develop numerous crowns and more flower stalks. Multiple rows are often arranged in groups of two, three, or four plants with a two-foot walkway between each group of rows. Plants are set about one foot apart in the multiple rows. The planting should be cultivated and hoed for the first two or three weeks; then the entire bed may be mulched (see page 16). Sawdust or wood chips or even layers of newspaper laid flat between plants may be used as soil mulch during the growing season. Apply sawdust or corncobs in a layer one to two inches deep. About four cubic yards are needed to cover 1,000 square feet.

Blossom Removal

Remove flower stems as early as they appear from newly set plants during the first summer. Allowing the fruit to develop during the first season delays root and runner development and reduces the crop the following year. Flowers that develop after July on everbearing and day-neutral cultivars should be left to produce a crop later in the fall.

Weeds

Cultivation and hand hoeing should begin soon after the plants are set. This practice will control weeds and make the soil more suitable for runner plants to take root. Repeated cultivation every seven to ten days is effective against weeds because weeds are easier to kill when they are small. Cultivation should be shallow around the plants to prevent injury to the roots.

Chemical herbicides can be used to control weeds, but they may be impractical for small gardens. Consult the nearest regional Extension office for up-to-date recommendations.

Fertilizing

Strawberry plants should be fertilized in early August with four to six pounds of ammonium nitrate (33 percent nitrogen) fertilizer per 1,000 square feet. This amount of nitrogen (one tablespoon spread in a narrow band about three inches from the crown of each plant) may also be applied about a month after planting if the plants are not vigorous. The August application may be broadcast over the rows but only when the foliage is dry. Brush the foliage with a broom or rake immediately after application to remove fertilizer particles; if left, the fertilizer particles may burn the leaves. Irrigate to carry fertilizers down into the root zone.

Be careful when applying fertilizer. Too much will cause excessive vegetative growth, reduce yields, increase losses from fruit and foliar diseases, and result in winter injury. Application of fertilizer during the spring of a fruiting year can produce soft berries and is not recommended.

Mulching

Strawberries should be mulched during the winter months to protect the plants from extreme cold as well as to reduce damage from frost heaving that occurs when the soil alternately freezes and thaws. Mulching also conserves soil moisture, keeps the berries clean, and provides better picking conditions. Use a loose organic material such as clean, seed-free wheat straw. The straw also can be used to cover the plants temporarily during cold nights in the spring to protect the flowers from frost injury.

Apply straw mulch after several frosts in the fall, but before the temperature drops below 20° F. This generally occurs between mid-November and mid-December in Illinois. If using heavy equipment, it is best to wait until the ground is frozen. Apply 100 to 150 pounds of straw per 1,000 square feet (two to four bales) three to four inches deep over the rows. If the straw blows off of the plants, it should be raked back on them (Fig. 9).

Remove part of the straw in the spring before new growth starts after the soil temperature remains at 40° F for at least three or four days. Put the excess straw between the rows.

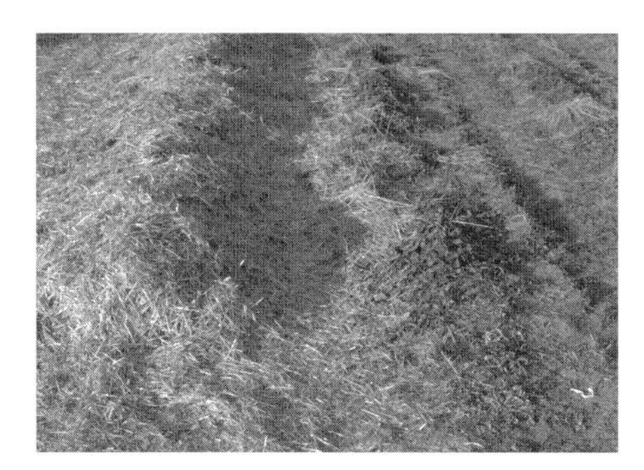

9 (left) This field of strawberries had been covered with straw (left) but it has blown off (right). To avoid plant loss, rake the straw back on the plants.

10a (above) In the spring, it is necessary to remove the straw to avoid suffocating the plants. Pull back the straw and look at the plants. If they are green and growing, remove the straw before they become pale.

10b (left) Strawberries will flower soon after the straw is removed (10a). Strawberry blossoms are very susceptible to frost. Frost-damaged blossoms can be seen the next moring as "black eyes" (Fig. 10c).

10c Strawberry flowers with frost-damaged centers called black eyes. These can be seen the morning after a frost.

Frost Control

Strawberry buds, blooms, and immature fruits are very susceptible to frost and freezing damage in the early spring (Figs. 10a, 10b, and10c). These losses can be lessened or prevented by covering the plants with straw or other insulating material or by careful and timely application of irrigation water. Irrigation water should be applied continuously when the temperature at the plant level reaches 34° F. Remember, because warm air rises and cold air sinks, the air temperature above a planting can be considerably higher than at the ground level. Sprinklers that put out a minimum quantity of water (0.1 to 0.3 inches per hour) should be used to prevent unnecessary flooding. A finely perforated plastic hose may also be used. Water will freeze on the plants and blossoms; the blossoms will not be injured as long as water is applied during the entire freezing period. This system is effective down to temperatures 25° F or lower. Once irrigation has begun, it must be continued through the night and into the next morning until all of the ice has melted from the plants.

Straw from between the rows may also be raked over the plants for protection. In small home patches, the placement of old blankets or sheets can often insulate the plants enough for them to survive frost. If this system is used, remove the covering (especially straw) during the day, and cover the plants on nights only when there is danger of frost. Spun bond material is available commercially to serve as a blanket over entire rows and will protect plantings down to temperatures of about 23° to 25° F.

Renewing the Patch (Renovation)

Properly managed strawberries will bear fruit more than one year. Usually a patch may be picked two to four years or more, but only good plantings should be maintained. Weedy or diseased plantings are best destroyed and replaced.

Immediately after the harvest is complete, the strawberries grown on the matted-row system should be renovated to achieve good production the next year. First, mow the old foliage with a power mower, cutting off the leaves about *one inch above the crowns.* Rake the leaves and other debris from the patch and burn, compost, or incorporate them in the soil. To avoid spreading leaf diseases, do not return the leaf mulch compost to the strawberry planting. Do not mow the leaves if renovation cannot be completed within a week to ten days after harvest. Broadcast ten to fifteen pounds of 10-

10-10 or 12-12-12 fertilizer per 1,000 square feet over the planting. Narrow the rows to six to twelve inches wide by spading, hoeing, or rototilling. Next, eliminate all weeds. If the remaining plants in the narrowed row are too crowded (closer than four to six inches apart), it may be advisable to remove some of them. If a herbicide is used, apply it carefully and as directed on the label. Irrigate thoroughly to encourage the plants to recover and make new runners for the next season's crop.

Day-neutral and everbearing strawberries are usually grown in a hill system (Fig. 8). Home gardeners who have limited space may grow these strawberries in terraced beds, pyramids, or barrels. These strawberries also make good edging plants or ground covers in the landscape. Sometimes they are grown as potted house plants or trained on "totem poles." If grown indoors, the plants must have good light, and the blossoms must be pollinated by hand to form well-shaped berries. Although these methods are not as productive as the conventional systems, they do have ornamental value.

TABLE 2.
Strawberry cultivars for Illinois, listed by season from earliest to latest within groups

		Disease resistance		
Strawberry type	**Fruit size**	**Red stele**	**Verticillium wilt**	**Region of adaptation**[*]
June bearing				
Earliglow	medium	R[a]	R	N,C,S
Annapolis	large	R	S	N
Honeoye	large	S	S	N,C,S
Delmarvel	large	R	R	N,C,S
Seneca	medium	S	S	N,C
Jewel	large	S	S	N,C
Kent	large	S	S	N,C
Allstar	very large	R	R	N,C,S
Day neutrals				
Tristar	medium	R	R	N,C,S
Tribute	medium	R	R	N,C,S

[a]R = resistant to this disease; S = susceptible to this disease.
[*]N = adapted to region north of Interstate 80; C = adapted to region between Interstate 80 and Interstate 70; S = adapted to region south of Interstate 70.

Brambles

The brambles include raspberries (below) and blackberries (left).

Raspberries

Raspberries ripen shortly after strawberries and are popular in all parts of Illinois. Plantings that are well cared for may produce crops for five years or more. Red, black, purple, and yellow fruit types are available. Red and yellow raspberries also have single- and double-cropping types. The double-cropping or everbearing cultivars bear one crop in the early summer and another crop in the fall. A careful pest-control program should be followed for all brambles (see page 61). Prior to planting, destroy any wild brambles growing around or near the new plantings because they harbor destructive insects and diseases.

TABLE 3.
Raspberry cultivars for Illinois, listed by season from earliest to latest within plant type

Raspberry type	Harvest season	Region of adaptation*	Notes
Cultivar			
Red raspberry			
Boyne	spring	N	old cultivar that is very winter-hardy
Latham	spring	N,C	old cultivar that is relatively thornless and dependable
Titan	spring	C,S	very large fruits, erect canes, very susceptible to Phytophthora
Heritage	spring + fall	N,C,S	widely adapted, erect, very dependable, firm berries, good flavor, susceptible to Phytophthora
Ruby	spring + fall	N,C,S	large fruits, very similar to Titan, but produces two crops per year
Southland	spring + fall	C,S	good quality fruit, adapted to southern regions
Yellow raspberry			
Goldie	spring + fall	N,C,S	reported to be a mutation of Heritage
Black raspberry			
Jewel	spring	N,C,S	largest fruits of the black raspberry cultivars
Haut	spring	C,S	very good flavor
Purple raspberry			
Brandywine	spring	N,C,S	large fruits that are good for jams or jellies
Royalty	spring	N,C,S	best-flavored purple, it can be picked when red to resemble the flavor of a red raspberry or left to darken and develop a flavor more like a black raspberry

*N = adapted to region north of Interstate 80; C = adapted to region between Interstate 80 and Interstate 70; S = adapted to region south of Interstate 70.

Plants and Plantings

Traditionally, one-year-old, No. 1 Grade plants have been best for establishing new plantings. Certified virus-free plants should be obtained. However, many commercial growers and homeowners are now planting tissue culture-derived brambles. These are small and require a little extra tending

during establishment, but with proper care, will outgrow and outyield conventionally propagated plants. Some nurseries offer "nursery-matured" tissue culture-derived plants that offer the same benefits as other tissue culture-derived plants but are easier to establish due to their larger size.

Raspberries are best planted in early spring (March or April). Prevent the plants from drying out in the field prior to planting by placing their roots in a bucket of water. After cutting off any broken roots, carefully spread the remaining roots in the planting hole and firm the soil over them. Set red raspberries at the same depth they were in the nursery; set black and purple raspberries about one inch deeper. Apply one or two quarts of water around each plant.

At planting time, conventionally propagated red raspberry plants should be cut back so that an eight- to twelve-inch "handle" protrudes from the soil after planting. The handle serves as a marker for the plant's location. Handles of black and purple raspberries should be cut off at ground level and removed and burned to prevent disease infestation. Tissue culture-derived plants need no pruning at planting.

11 Fertilizer application should be made around the plant at the drip line.

Plant Spacing and Support

Raspberries may be grown in hills or in hedgerows. The plant spacing depends on the system of training to be used. (see Table 1 on the inside front cover and the discussion of training systems on page 26). Red raspberries spread by root suckers and naturally form a hedgerow. Black and purple raspberries seldom spread by root suckers and will remain as individual plants or hills.

Fertilizing

For maximum production, fertilizer should be applied prior to planting. Ten to fourteen days after planting, apply two ounces of 5-10-5 fertilizer around each plant.

In the second and subsequent years, the plants should be fertilized with 10-10-10 or equivalent fertilizer at a rate of fifteen to twenty pounds per 1,000 square feet broadcast along the hedgerow or about one-half cup spread around each plant in the hill system(Fig. 11). Apply fertilizer in early spring before new growth begins.

Dry animal manures may also be used to fertilize established raspberry plants. In the spring before new growth begins, apply 300 to 400 pounds of cow manure per 1,000 square feet or 100 to 200 pounds of poultry manure per 1,000 square feet.

Do not apply fertilizer during the late summer or early fall. Such applications may injure the plants or stimulate soft, succulent growth that is very susceptible to winter injury.

Mulching

Generally, raspberries should be cultivated shallowly during the early part of the first summer to suppress weeds. After the plants are established, a light organic-matter mulch may be applied. Mulched raspberries grow better, produce more, and have larger berries. A light cover of wheat straw is a good mulch. To suppress weed growth, mulch should be two to four inches deep. Too deep of a mulch will suppress cane emergence as well as weeds. Heavy mulches such as leaves and corncobs will suppress cane growth. The mulch should be renewed annually if needed. If turf is used between rows, it is best to keep it closely mowed and out of the row.

Mulched plantings will require extra nitrogen over the first two years. Apply double the amounts of fertilizer recommended earlier. After two years, the amount of nitrogen fertilizer applied may be reduced by half because the decomposing mulch will begin to release fertilizer nutrients to the plants.

Blackberries

Blackberries are well suited to the home fruit garden in the southern half of Illinois. Most blackberry cultivars are too susceptible to winter damage for dependable production in northern Illinois. However, one cultivar is suitable for culture in most of northern Illinois (see Tables 1 and 4 on the inside front cover and page 33). Blackberries are classified as thorny, thornless, erect, semi-erect, or trailing. Erect, semi-erect, and trailing type blackberries are all available, but each requires different culture. The trailing and semi-erect cultivars, including most of the thornless blackberries, require a physical support and are not very winterhardy. Thus, the culture of most thornless blackberries in Illinois requires special care. Plant breeders are working to develop hardy thornless erect cultivars.

Planting and Spacing

Blackberries are best planted in early spring, using the same care as when planting raspberries. Spacing will depend on the trellis and training system to be used (Figs. 14 and 15 on pages 27 and 28). Most erect blackberry cultivars can be grown without physical supports and are spaced two to four feet apart in rows eight to ten feet apart. Semi-erect and trailing blackberries are planted eight to ten feet apart in rows eight to ten feet apart.

Establish your planting from virus-indexed (certified) plants. Some nurseries can supply tissue culture-derived plants. Tissue culture-derived plants grow more vigorously and

12 Tissue culture-derived plant.

become established more quickly than conventionally propagated plants. Tissue culture-derived plants are supplied in three-inch root plugs (Fig. 12). When these small plugs are planted, the roots should be barely covered with soil, whereas conventionally propagated plants are set at the same depth as they were when planted in the nursery. The tops of conventionally propagated plants should be cut back to six inches; tissue culture-derived plants need no pruning at this time. Cultivation, mulching, and fertilizer are the same as for raspberries (see pages 22–24).

Winter Protection

Most of the semi-erect and trailing thornless blackberry cultivars are not very winterhardy and are suggested only for southern Illinois. However, with special protection, they might be grown in northern areas, though some winter injury should be expected even then. After they become dormant in the fall, the canes can be protected by removing them from the trellis support and completely covering them with soil or straw. Any exposed canes can be damaged by cold air. When the danger of severe cold is past in the spring, uncover the canes, remove broken or dead canes, dormant-prune the other canes, and tie the remaining canes to the support.

Sterility Problems

Sterility in some blackberry cultivars has been a problem. Affected plants generally grow vigorously and bloom profusely, but set only a few malformed berries. If such plants occur in your garden, they should be destroyed immediately—roots as well as stems. The cultivars listed in Table 4 on page 33 have been observed to bear consistently in several locations in Illinois. However, their resistance to sterility problems is unknown.

Training and Pruning Brambles

Growth & Fruiting Habits

To prune the brambles properly, it is essential to understand their growth and fruiting habits. Brambles bear fruit on biennial canes that live for two years. The roots and crowns are perennial. All brambles send up new shoots during each growing season from the crown. The first year, these new shoots are called primocanes. Erect blackberries and red or yellow raspberries develop primocanes from both the crown and roots. These primocane shoots, regardless of origin, grow vigorously during the summer, initiate flower buds in the fall, overwinter, and bear fruit the following season. In the second season, the shoots are called floricanes. The fruit is borne on leafy shoots arising from lateral buds on the floricanes. After harvest, the floricanes gradually dry up and die. Meanwhile, new primocane shoots are developing to repeat the cycle, thus providing fruiting canes each year.

13 Black and purple raspberries and erect blackberries are pinched back three to four inches after the primocane has reached the desired height in the summer. This results in the development of lateral shoots shown in Fig. 17 on page 31.

Some brambles fruit twice on the same cane. These are called everbearing, fall bearing, double cropping, or primocane flowering. At present, only red and yellow raspberries are available as primocane-bearing cultivars. The primocanes bear a crop at their tips in the fall and again the next season as floricanes further down on the canes, after which the canes die.

Variation in the development of lateral branches must be considered in order to properly prune brambles. Black raspberries, purple raspberries, and erect blackberries develop strong lateral branches when the new shoot tips are cut back early in the growing season. This characteristic is encouraged by the practice of pinching, or summer topping, which consists of snapping off or cutting with shears the top three or four inches of the new shoots (Fig. 13). This system has no advantage for red and yellow raspberries, semi-erect blackberries, and trailing blackberries. If red or yellow raspberries are topped, weak laterals develop. These break easily and are likely to be winterkilled.

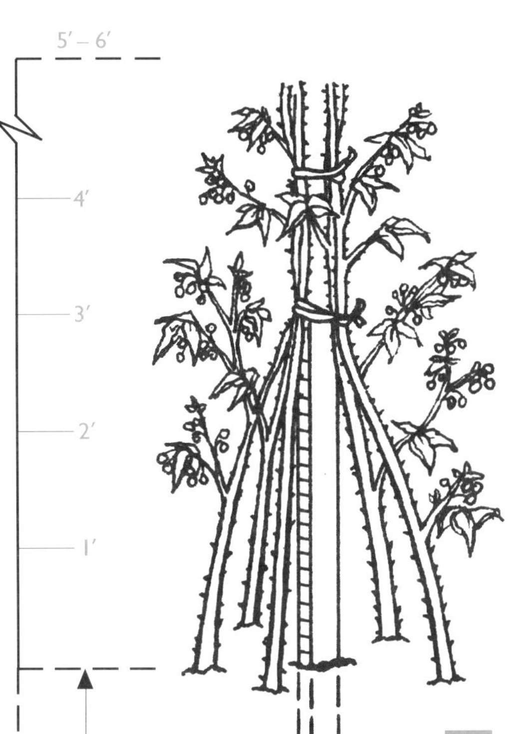

14 These raspberries are growing on a hill system; their canes are trained to a pole. An example of a mature plant trained to this system is shown in Fig. 19 on page 32.

Supports

Although most of the brambles can be grown without support, some type of support or trellis is desirable to produce high-quality fruit. Trailing and semi-erect blackberries must be supported. Fruit on supported plants is clean, and a minimum of the crop is lost due to cane breakage by wind, cultivation, and picking. Supports also facilitate harvesting and other cultural practices and in the long run will pay for themselves. Trellised plantings can usually be kept attractive in the home landscape.

Many types of trellises and methods of training are in use by bramble growers. The following are some of the simplest methods of satisfactory trellis construction.

The hill system of culture (Fig. 14) may be used for any of the bramble fruits. A single stake, two to four inches in diameter, driven into the ground, supports the plant. Five to eight canes are tied in one or more places to the stake in the spring following dormant pruning. The plants are usually set four to six feet apart each way.

15 Two types of trellises are shown. The horizontal trellis (top) is suitable for red raspberries and requires minimum tying of canes. The two-wire vertical trellis (bottom) allows canes to be tied to it.

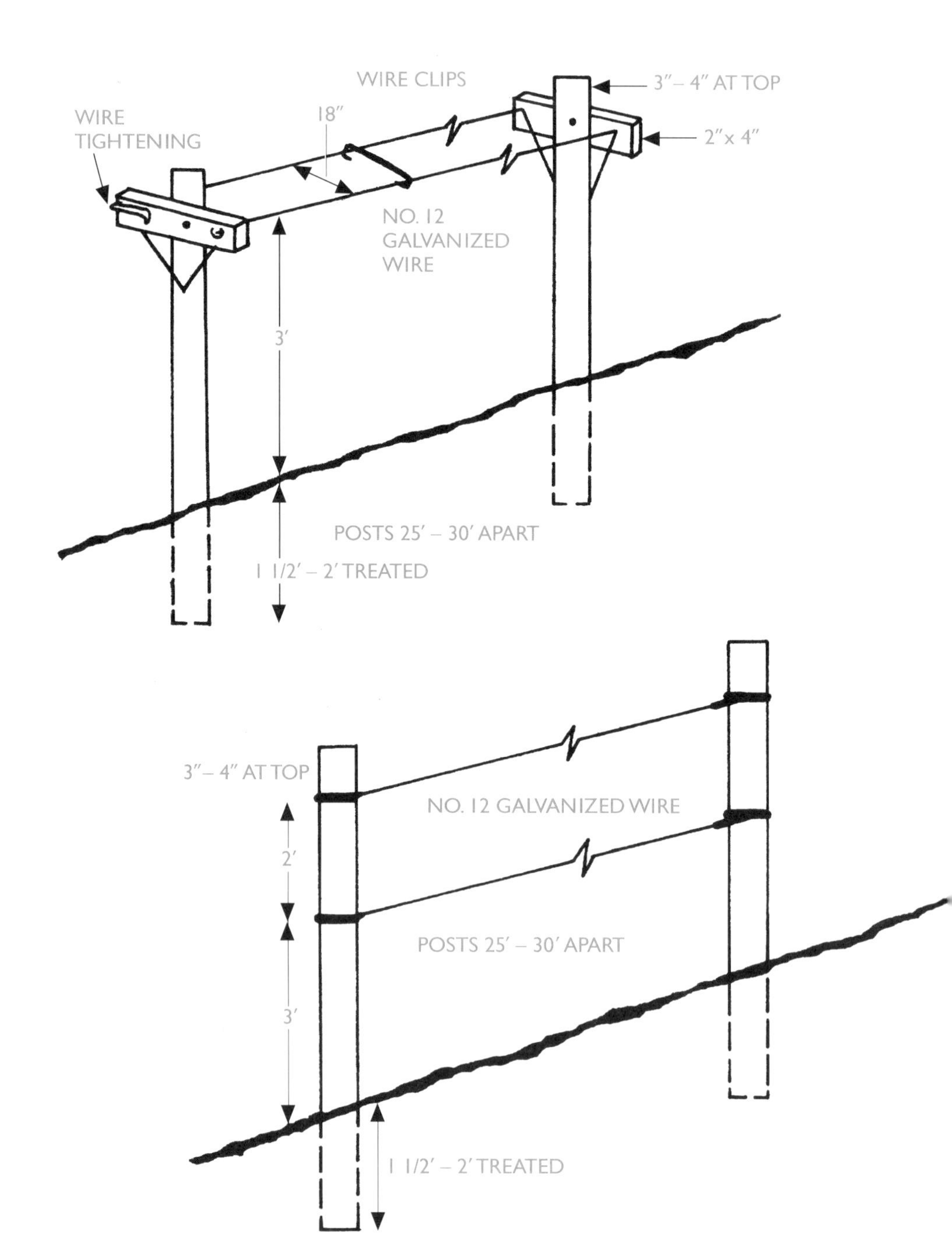

Wire trellising systems (Fig. 15) are more practical for large plantings. Posts for building a trellis may be either metal or wood. If wood posts are used, buy treated posts for extended usefulness. Posts should be spaced twenty-five to thirty feet apart in the row, with the end posts braced or anchored. A nine- to twelve-gauge galvanized wire is adequate for the trellis.

For raspberries and erect blackberries, the most common system of training is the horizontal trellis. The plants form a solid hedgerow that is kept about eighteen inches wide. Each spring the canes within the row are thinned so that they are at least six to twelve inches apart. The canes are supported between the trellis wires and do not have to be tied when using this system. Wire clips can be used to keep the wires from spreading between the posts.

The vertical trellis is used with trailing blackberries and semi-erect thornless blackberries. This system allows excellent weed, disease, and insect control. Plants are maintained in a narrow row and are tied to the trellis wires. Harvesting of these plants is also easier when they are secured to a trellis.

Pruning Red and Yellow Raspberries

Red and yellow raspberries should be pruned two times per year, regardless of the type of training system used. The first pruning in the early spring reduces cane number by thinning. The second pruning should occur as soon as possible after fruiting. *New shoots of red and yellow raspberries should not be summer topped.*

16 Red raspberry plant before (left) and after (right) dormant pruning.

Spring pruning should be done early, before the buds begin to swell, but after danger of severe cold is past. All short and weak canes should be removed, and the remaining vigorous canes should be thinned to five to eight canes per stake in the hill system or spaced six to twelve inches apart in other systems. Only the largest canes should be saved, as these are the most fruitful. If too tall, cut the remaining canes back to five or six feet from the ground level. Canes shorter than five to six feet do not need to be cut back unless winter injury extends below this height. If no support is provided, cut the canes back to about three to four feet, regardless of their original height (Fig. 16, page 29).

Fruited canes should be removed any time after harvest since they die soon after fruiting. Although this pruning may be delayed until the following spring, cutting these canes off at the ground level as soon as possible after harvest is preferred. Removal at this time encourages growth of the new shoots and reduces disease and insect problems. Prunings are removed from the planting. Prunings may be burned or chipped and composted, but the compost should not be returned to the bramble planting because of insects and diseases that may be harbored in the old tissue.

The canes of primocane-bearing varieties are pruned the same way as ordinary varieties: in the spring (thinning) and after the summer harvest (removing canes that bore the crop). The new shoots will bear fruit in the fall at the tip and bear again the next spring further down the cane, *so do not remove these canes after the fall harvest if a spring crop is desired.*

However, since the primocane-bearing types produce abundantly on first-year growth, many growers harvest only once. All of the canes are mowed as close to the ground as possible in the spring before growth begins. This practice eliminates all labor for hand pruning and the problems associated with winter injury to canes. 'Heritage' is the best-suited cultivar for this fall-bearing system. In addition, 'Heritage' is very erect and can be grown with little or no trellising—another labor- and money-saving advantage.

Pruning Black Raspberries, Purple Raspberries, and Erect Blackberries

These brambles should be pruned three times per year—during the summer, after harvest, and early spring.

Summer pruning is an essential step in the production of these brambles. All new shoots should be pinched back three to four inches once they have grown to the desired height (Fig. 13, page 26). If grown without supports, black raspberries are pinched when the shoots are about twenty-four inches high; purple raspberries and erect blackberries are pinched when they reach thirty to thirty-six inches. If the brambles are grown with supports, the shoots can grow six to eight inches more before pinching. Some of the thorny blackberry cultivars from Arkansas (Table 4 on page 33) are strong enough to be self-supporting at heights of four to five feet. These cultivars can be pinched back when they attain a height of about forty-eight

17 Black raspberry plant is shown before (left) and after (right) pruning. Purple raspberries and erect blackberries are pruned in a similar manner.

inches. The pinching operation usually coincides with harvest. The plantings should be examined several times because primocanes develop over a period of several weeks.

As with red and yellow raspberries, fruited canes of black and purple raspberries and erect blackberries should be removed after harvest. Dormant pruning (Fig. 17) is best done in the late winter to early spring before the buds begin to swell. All weak canes should be cut out at the ground level, leaving four or five of the most vigorous canes (at least one-half-inch diameter) per plant. The lateral branches on the remaining canes are thinned out and shortened. All broken, weak (usually one-fourth-inch diameter or less), and dead laterals are removed. The laterals of black raspberries are shortened to eight to ten inches of growth, or eight to twelve buds per lateral. Purple raspberries and erect blackberries are more vigorous; thus, the laterals may be left twelve to eighteen inches long, or with about fifteen buds per lateral.

Erect blackberries send up root suckers in addition to the new canes that grow from the crown. If all of these have been allowed to grow during the previous season, they should be thinned out during the dormant pruning process to leave about three to four canes per linear foot of row. A better practice is to remove the excess suckers during the summer, leaving only those desired for fruiting next season. The suckers should be pulled out, if possible, since they do not regrow as quickly as those that are cut off at ground level.

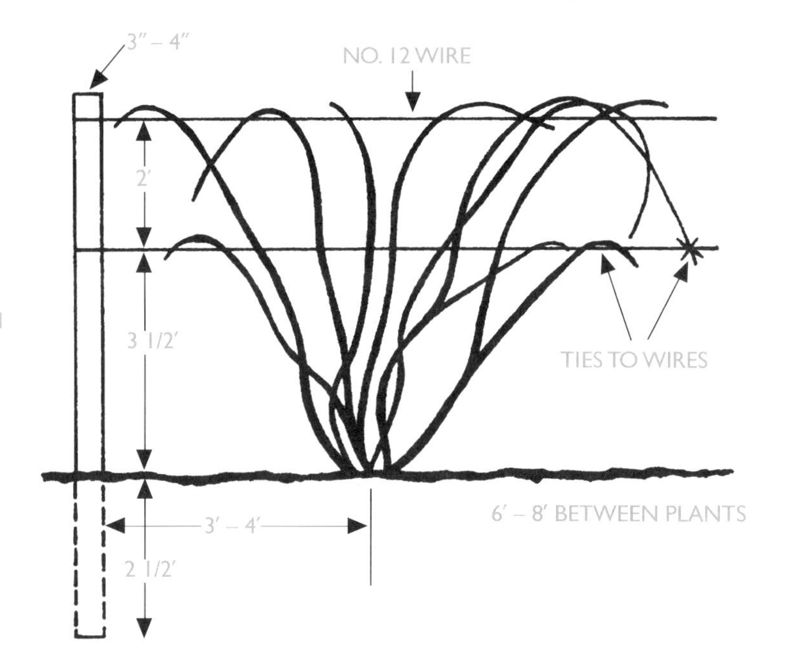

18 Dormant pruning and training of trailing and semi-erect blackberries to a two-wire vertical trellis.

Pruning Trailing and Semi-Erect Blackberries

These brambles are pruned twice, once in early spring and again after harvest. Pinching the new shoots as they develop is not practical and may lead to winter damage. The tips of thornless trailing blackberry canes are the first parts to show winter kill. Therefore, if you leave the canes long as the plants go into winter, the lower part of the cane may still be alive in the spring.

Spring pruning of these brambles consists simply of selecting the best canes or cane parts that have survived the winter, tying them to a trellis, and

19 A trailing blackberry plant trained and pruned to the hill system. Note the new primocanes growing from the base of the plant.

removing damaged and diseased canes (Figs. 18 and 19). Select the best six to eight canes. If stakes are used for support, the canes can be wrapped around the stake, tied at two or three places, and cut off to the height of the stakes. On a horizontal trellis, the canes are wound around the wires and tied. On a vertical trellis, the canes are tied to the wires without being wound.

Remove all fruiting canes after harvest at ground level.

TABLE 4.
Blackberry cultivars for Illinois, listed by season from earliest to latest within plant type

Blackberry type / Cultivar	Harvest season	Region of adaptation*	Notes
Thornless			
Hull Thornless	late summer	S	large fruit, good flavor, not very winterhardy; plants are semi-erect and need to be trellised
Chester Thornless	late summer, harvest can last up to 6 weeks	C,S	large, round fruit; fruit is firm and could be transported to local markets without leaking; plants are semi-erect and need to be trellised
Triple Crown	mid-summer; earlier than Chester	S	Newest semi-erect thornless; suitable for trial only
Arapahoe	July	S	erect thornless suitable for southern Illinois
Thorny			
Choctaw	very early	S	good flavor, erect growth habit
Cheyenne		S	good flavor, erect growth habit
Shawnee		S	very large fruits, good flavor, erect growth habit
Illini Hardy	long harvest season	N,C,S	dependable moderate production of good quality fruits

*N = adapted to region north of Interstate 80; C = adapted to region between Interstate 80 and Interstate 70; S = adapted to region south of Interstate 70.

Blueberries

Blueberries are delicious when eaten fresh, are tasty in pies and muffins, and are easily frozen. However, this fruit is not commonly grown in Illinois by most home gardeners. Blueberries have very exacting soil and cultural requirements, but, if properly handled, can be grown successfully. Soils and locations that naturally provide optimal growing conditions are very limited in Illinois, so careful attention must be given to cultural details. Blueberries grow and yield best when grown in full sun but will produce good yields in partial shade.

TABLE 5.
Blueberry cultivars for Illinois, listed by season from earliest to latest within plant type

Cultivar	Region of adaptation*	Notes
Collins	N,C,S	dependable early cultivar
Patriot	N,C,S	new early cultivar
Bluejay	N,C,S	very productive
Bluecrop	N,C,S	a dependable, widely adapted, standard cultivar
Herbert	N,C,S	a very dark-colored fruit; best-flavored blueberry available
Nelson	N,C,S	productive, good flavor, light color
Elliott	N,C	very late, productive, extends the blueberry season into late August or September

*N = adapted to region north of Interstate 80; C = adapted to region between Interstate 80 and Interstate 70; S = adapted to region south of Interstate 70.

Soil

Blueberries require an acidic soil relatively high in organic matter. A soil pH of 4.8 to 5.2 is best for optimal growth. The soil should be tested prior to planting. Check with the regional Extension office or refer to the telephone listings for the nearest soil-testing services. The pH measurement refers to the relative acidity or basicity (alkalinity)of soil. The pH level is measured in numbers 1 to 14. A measurement of 7 indicates neutral; 1 to 7 indicates acidic; and 7 to 14 indicates basic.

If your soil is between 5.2 and 6.2, you can make it more acidic by adding a mixture of half acid sphagnum peat moss and half top soil. Use this soil-peat mixture in a two-foot-diameter planting hole at least twelve inches deep. If your soil pH is above 6.2, individual plants can be grown in tubs buried in the garden. Halves of fifty-five-gallon drums, with drainage holes cut in the bottom, are suitable. First, burn out any residues that might be injurious to the plants. Bury each drum in a sunny area, and leave one to two inches of the rim above ground level. Fill the tub with an acidic soil (pH 4.8 to 5.2) that is high in organic matter. Approximately 10 to 15 percent (volume: volume) of sphagnum peat moss is good for this purpose. Set one blueberry plant in each tub.

Irrigation

Blueberries are shallow rooted and grow best where the water table is fourteen to twenty-two inches below the soil surface. Because of this water requirement, blueberry plants must be irrigated in most parts of Illinois. About an inch of water per week is sufficient. Do not attempt to grow blueberries unless you can supply the necessary water when rainfall is not adequate. In addition, the soil must drain freely; blueberry plants do not tolerate standing water.

Mulching

Because blueberries are very subject to drought injury, mulching with organic matter is recommended in most Illinois soils, in addition to irrigation. The mulch also supplies organic matter and helps control weeds (Fig. 20). Although some mulches are better than others, the use of any mulch is better than no mulch at all, especially on the heavy clay soils found in much of Illinois. A deep mulch of sawdust, chopped cornstalks, or leaves is best, but other organic matter such as wood chips or straw is also useful. Apply four to six inches of mulch soon after the plants are set.

Maintain the mulch by adding two to four inches as needed. In rows, the mulch should be maintained three to four feet wide. The aisles between rows can be kept in mowed turf. On small home plantings, the entire blueberry area can be mulched to control weeds. Research at the University of Illinois has demonstrated that plants grown with a mulch produce significantly larger fruit yields than plants grown

20 The use of mulch is important for blueberry culture in Illinois. Both plants shown here are six years old. The one in front has had no mulch; the one in back has been grown with cornstalk mulch.

without mulch, even when both plantings are irrigated equally.

Planting and Spacing

Buy container-grown, two-year-old plants of medium-to-large size. If possible, avoid using plants older than three years; their extra cost is not justified, and they may be cull plants that were too weak to be sold at a younger age.

For each plant, remove one bushel of soil and mix it with one bushel of sphagnum moss (if needed). Put half of the mixture in the hole, set the plant, and fill the hole with the remaining mixture.

The plants should be set in the hole at the same depth they were growing while in the nursery, not deeper. Do not let the roots dry out during planting. Carefully spread the roots and firm the soil or soil-peat mixture over them. Blueberry plants should be spaced six to eight feet apart in rows eight to twelve feet apart. Water thoroughly after planting. Remove one-half to two-thirds of the cane growth at planting time.

A vigorous three-year-old blueberry bush heavily laden with fruit.

Removing Flowers

Remove all flowers during the first and second year. Do not permit berries to develop because it will restrict shoot and root growth.

Fertilizing

Do not apply fertilizer until four weeks after planting. Then, sprinkle one ounce of ammonium sulfate in a circular band twelve to eighteen inches from the base of each plant. The ammonium sulfate will provide nitrogen for growth and help maintain the acidic soil conditions required by blueberries. *Do not use aluminum sulfate*: it is toxic to blueberries.

In the spring of the following year, apply two ounces of ammonium sulfate per plant in late March or early April before the buds begin to swell. Increase this amount each year by one ounce until a total of eight ounces per plant is reached. After that, each plant should receive eight ounces annually. A complete fertilizer, such as 10-6-4 or 10-10-10 analysis, can be used at double the above rates if growth is not vigorous and the pH remains between 4.5 and 5.5. Check the soil pH every one or two years. Fertilizers prepared for acidic soil-loving azaleas and rhododendrons also can be used at the same rates as ammonium sulfate.

Iron deficiency of blueberries can be a problem. Symptoms are yellowing and mottling of the young leaves. Iron chlorosis in blueberries is usually an indication that the soil pH is too high, thus making iron unavailable to the plant. This may be temporarily corrected by applying iron chelate to the soil or by spraying it on the foliage in amounts recommended by the manufacturer. However, most Illinois soils contain adequate amounts of iron. The soil pH should be tested and amended periodically to maintain it at the proper level.

Pruning

Pruning is generally not needed until the third year after planting if growth is normal. In early spring before growth begins, remove dead or injured branches, short or stubby branches near the ground, and old stems of low vigor. Leave vigorous branches unpruned (Fig. 21). After the plants are five to seven years old, it is important to remove some of the old canes each year. In general, it is a good idea to maintain five to seven older canes, as well as one or two new canes. Once these numbers are attained, one or two older canes can be removed annually, and one or two new canes can be left to replace it. This practice allows vigorous younger canes to develop and facilitates complete bush renewal over a five-to seven-year period.

Pruning also increases the size of the berries and promotes earlier ripening. If the plants have formed an unusually heavy load of fruit buds, the tips of the fruiting branches can be cut back to leave four to six fruit buds. Although

21 A four- or five-year-old blueberry bush is shown. Left, before pruning. Right, the same bush after removal of weak and unproductive growth.

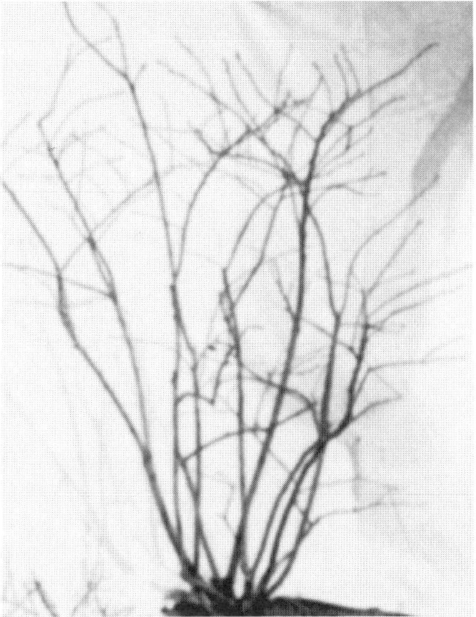

22 Blueberries with net for bird protection.

this reduces yields slightly, the berries are appreciably larger. The fruit buds are easily distinguished in the spring because they are large, round, plump buds. Leaf buds are smaller, thinner, and sharply pointed.

Under good growing conditions, vigorous shoots may rapidly develop and grow several feet tall. If the tips are cut back before August 1, the canes usually develop strong lateral branches that bear fruit the following spring. When the shoot is four to five feet high, three or four inches should be removed.

Vertebrate Pests

Birds are a serious pest in blueberry plantings. As the berries ripen, the bushes can be covered with protective netting (Fig. 22), either over individual plants or whole plantings. Although the use of noise deterrents may help to keep birds away from the planting, the usefulness of these deterrents is limited because birds become accustomed to the noises and learn to ignore them. There are so-called "hawk eye" balloons that can be flown in a planting. Birds flying by recognize the large eyelike markings on the balloon, apparently interpret it as a predator bird, and fly on. However, hopping and low-flying birds such as robins and brown thrashers are not deterred by such scare devices.

Other animals also pose problems. Rabbits and deer may eat twigs and branches when the ground is covered with snow. Fencing is warranted when this problem is severe. Voles can also be a serious problem in mulched plantings.

SOURCES FOR BIRD NETTING AND SCARE DEVICES

BIRD-X, INC.
300 North Elizabeth
Chicago, IL 60607

FORESTRY SUPPLIERS, INC.
P.O. Box 8397
Jackson, MS 39284-8397

GEMPLER'S
211 Blue Mound Road
Mt. Horeb, WI 53572

HUMMERT INTERNATIONAL
4500 Earth City Expressway
Earth City, MO 63045

INDIANA BERRY AND PLANT CO.
5218 West 500 South
Huntingburg, IN 47542

Currants & Gooseberries

Currants and gooseberries are very hardy and easy to grow. They will be discussed together because their cultural practices are similar. These fruits are little known in Illinois. They were prohibited from culture for many years in this state because they can serve as an alternate host for the white pine blister rust disease. However, at the present time, there is little blister rust in Illinois, and the state restrictions on growing these plants have been removed.

Currants (page 42) and gooseberries (above).

TABLE 6.
Currant and gooseberry cultivars for Illinois, listed by season from earliest to latest within plant type

Fruit type Cultivar	Region of adaptation*	Notes
Red Currants		
Red Lake	N,C,S	an old dependable cultivar
Cherry	N,C,S	an old cultivar
Black Currants		
Consort	N,C,S	immune to white pine blister rust
Crandall	N,C,S	not susceptible to white pine blister rust
White Currants		
White Imperial	N,C,S	mild flavor that resembles the red currant
Gooseberries		
Pixwell	N,C,S	light green fruit, plant has few thorns
Poorman	N,C,S	large red fruit, mildew resistant
Welcome	N,C,S	dull red fruit, plant has few thorns
Captivator	N,C,S	essentially thornless
Black currant X gooseberry hybrid		
Jostaberry	N,C,S	large fruit with a flavor similar to but milder than the black currant parent

*N = adapted to region north of Interstate 80; C = adapted to region between Interstate 80 and Interstate 70; S = adapted to region south of Interstate 70.

Location

Currants and gooseberries grow best in cool, moist, and partially shaded locations. The plants prefer a loam soil with 3 to 5 percent organic matter.

Planting and Spacing

Plants can be set in either fall or spring, but obtaining plants in the fall may be a problem. Spring planting is quite satisfactory if done early before the buds begin to grow. Vigorous, well-rooted, one-year-old plants are best. Damaged roots should be pruned off; the top should be cut back to ten inches. Plants should be set a little deeper than were grown in the nursery to encourage a bush form to develop. Plants should be spaced four feet apart in rows six to eight feet apart.

Mulching and Fertilizing

Currants and gooseberries are heavy feeders but have rather shallow root systems. An annual application of rotted manure or decomposed compost is ideal for these plants. Well-composted strawy manure may be applied each fall (November) and maintained four to six inches deep to provide a soil mulch. Sawdust, straw, lawn clippings, compost, or similar materials may be used for mulch. If the plants lack vigor, four ounces of 10-10-10 fertilizer should be applied per plant in early spring before growth starts. This amount should be doubled the first year the plants are mulched.

Pruning

Currants and gooseberries require annual pruning for maximum production. The fruits develop from buds at the base of one-year wood and from spurs on older wood. The older wood becomes progressively less fruitful, and canes older than three years are usually unproductive. Pruning consists mainly of selecting the proper type of fruiting wood and removing the unproductive older canes.

Prune in the early spring while the plants are still dormant. After the first year, remove the weaker shoots, leaving six to eight strong canes. The second year, remove all but four or five of the two-year-old canes and four or five of the one-year-old canes. On the third and subsequent years, leave four or five of the three-year-old canes, four or five of the two-year-old canes, and four or five of the one-year-old canes per plant. When pruning, remove canes that tend to lie on the ground; remove weak canes in the center of the bush to prevent the bush from becoming too dense.

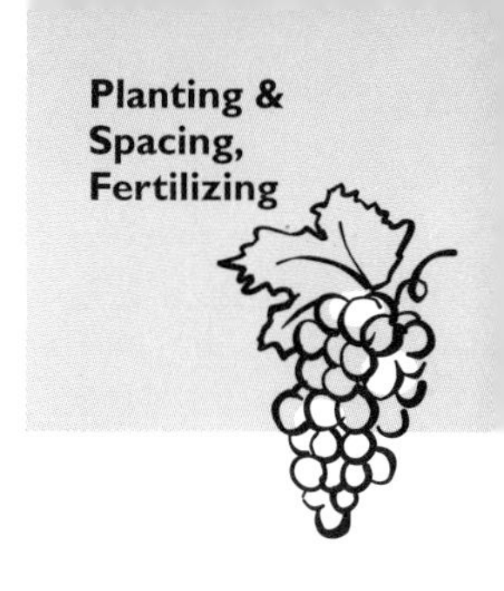

Grapes

Grapes are a popular fruit for home gardens. By selecting early, midseason, and late cultivars, the home gardener can extend the harvest season from mid-August to mid-October. Grapes offer a wide range of flavors and can be eaten fresh or processed into wine, juice, and jelly. The trellises and arbors on which they are grown can function as attractive shaded areas in the landscape and as screens to block undesirable views. In recent years, French hybrid grapes have created interest in making wine. Several moderately hardy seedless cultivars are suitable for table use.

Planting and Spacing

Grape vines should be planted in the early spring as soon as the soil can be prepared. Cut off any long or broken roots so that the remaining roots can be spread evenly in the planting hole. Set the plant slightly deeper than the depth it grew in the nursery. After planting, prune as directed on page 50. Space the plants eight to ten feet apart in the row and not less than eight feet between rows. If grafted vines are used for phylloxera resistance, do not set the graft union below the soil level.

Fertilizing

Have your soil tested prior to planting. Before planting, add nutrients on the basis of this test. If the selected cultivars are vigorous and produce excessive growth, *do not apply fertilizer.* The use of excessive fertilizer can encourage rank growth that does not mature properly, is unproductive, and is likely to be damaged during cold winters. Canes that grow moderately tend to mature early and produce better crops than excessively vigorous canes.

TABLE 7.
Grape cultivars for Illinois, listed by season from earliest to latest within plant type

Grape type Cultivar	Fruit color	Region of adaptation*	Uses
American grapes			
Ontario	white	C,S	very sweet dessert grape
Buffalo	blue	C,S	juice, pies, jelly
Niagara	white	N,C,S	large yields, juice
Cayuga White	white	C,S	use for wine making
Steuben	red-blue	N,C	dessert and juice
Swenson Red	red	N,C	good flavor dessert grape
Fredonia	blue	N,C,S	juice and pies
Delaware	red	N,C,S	juice and wine
Concord	blue	N,C,S	juice, jellies, grape pudding
Catawba	red	S	latest grape in Illinois, juice and wine
Seedless grapes			
Canadice	red	C,S	small dessert grapes similar to Delaware
Himrod	white	S	makes a good frozen grape
Reliance	pink-red	N,C	good flavor, the hardiest of the better seedless grapes. Makes a good juice and can be dried as sticky raisins.
Glenora	dark blue	C,S	high quality dessert grape
Einset	red	S	new cultivar for trial
Remailly	white	S	new cultivar for trial
Concord Seedless	blue	N,C,S	hardy, small fruit, flavor similar to Concord
French hybrids (wine grapes)			
Vignoles	white	C,S	wine grape, vigorous
Marechal Foch	blue	N,C,S	hardy, productive, makes good wine
Seyval	white	C,S	wine grape, productive
DeChaunac	blue	C,S	wine grape, some disease resistance
Chancellor	blue	S	wine grape, frost tolerant
Chambourcin	blue	S	good wine grape
Baco Noir	blue	C,S	highly productive, resistant to downy mildew

*N = adapted to region north of Interstate 80; C = adapted to region between Interstate 80 and Interstate 70; S = adapted to region south of Interstate 70.

If the vines are not vigorous growers, a commercial fertilizer may be used. Apply two ounces of a 10-10-10 fertilizer (or equivalent) around each vine shortly after planting. In the early spring of the second year, apply four ounces; the third year, eight ounces; and thereafter, one pound per plant. Distribute the fertilizer around the plant. If the soil is high in potassium and phosphorus, 33 percent nitrogen fertilizer (ammonium nitrate [NH_4NO_3]) can be substituted at one-third the above rates.

Mulching

Cultivate the young grape planting for the first year. Do not cultivate deeper than three inches around the plants because the roots are rather shallow.

After the vines are established, the grapes can be mulched with straw, sawdust, leaves, or compost. Caution must be used when mulching grape vines on heavy, wet, or highly fertile soils. Mulching may encourage overly vigorous canes that will fail to mature properly in the fall, leading to winter damage.

Supports

The first season of growth can be supported by a one-inch by one-inch wooden stake. A trellis or arbor should be constructed before the second growing season. The grape planting is more or less permanent, and the support trellis should be built to last twenty or more years. The structure, therefore, must be strong enough to bear the weight of mature vines as well as a full crop.

The most common training system is the four-cane Kniffin system (described later). A trellis for this system consists of two wires, approximately three feet apart vertically, supported by posts set about twenty to thirty feet apart in the row. Nine- or ten-gauge galvanized wire is suggested. The lower wire should be three feet above the ground, and the top wire about five and a half to six feet above the ground. Durable wood posts (cedar, locust, white oak, osage orange, or green-tinted CCA-treated timber rated for ground use) should be a minimum of three inches in diameter at the top and eight to eight and a half feet long. The posts should be set two and a half to three feet in the ground; for long rows, heavier posts should be used for the ends of the trellis. The end posts should be five to six inches or more in diameter at the top, and they should be nine feet long so that they can be set a full three feet deep. The end posts should be well braced to keep the trellis wires from sagging (Fig. 23, next page). Metal posts also may be used for supporting trellis wires.

Grapes can be trained satisfactorily on latticed arbors, fences, or other suitable structures. For some homeowners, the shading provided by grape vines growing over arbors may be as important as the fruit crop.

23 Two examples of well-braced grape trellises (above and upper right).

Pruning

Grape vines produce the greatest amount of high-quality fruit when they have moderate vigor. Weak vines do not have enough strength to produce normal crops. Very vigorous vines spend too much energy on cane growth and not enough energy on fruit production. As discussed earlier, one way to modify vigor is to reduce the fertilizer rate. However, the best way to adjust the fruit load and maintain proper vigor is through annual pruning.

Grapes flower and produce fruit only on one-year-old canes. The most productive wood is on the six to eight buds closest to the base of the cane. The most productive canes have moderate vigor and are about the diameter of a common yellow pencil.

Therefore, the goal of pruning is to encourage vigorous new canes to develop, to eliminate unproductive old canes, to train fruiting canes, and to limit the number of buds on the vine. Most gardeners do not prune severely enough. Proper pruning often results in the removal of 80 to 90 percent of the wood.

Prune after the coldest part of the winter is past, but before the buds begin to swell. February and early March are usually the best times in Illinois. Pruning during the summer is not recommended except to control excessive growth.

Regardless of the training system used, American types and French hybrids should be pruned annually using one of the following methods.

Pruning: American Types

The "30 + 10" rule is an excellent guide for balanced pruning of 'Concord' and other American cultivars. For each vine, thirty buds are left for the first pound of prunings removed and ten buds are left for each additional pound of prunings. For example, a one-pound vine (one pound of prunings removed annually) should have thirty buds left; a three-pound vine should have fifty buds left. Cultivars vary, but one- to three-pound vines are the most productive.

A simpler method of pruning American cultivars is to leave forty-five to sixty buds on vigorous plants and thirty to forty buds on low vigorous vines regardless of training method.

Pruning: French Hybrids

French hybrids tend to set more grapes per bud than American types; therefore, fewer buds should be left. For these hybrids, a balanced pruning system using a "20 + 5" rule is suggested, that is, twenty buds for the first pound of prunings and five buds for each additional pound of prunings.

A simpler method of pruning for French hybrids is to leave twenty-five to thirty-five buds on vigorous vines and twenty to twenty-five buds on the less vigorous plants, regardless of training method. In general, short-fruiting cane spurs with two to seven buds are preferred for the French hybrids. Some shoot thinning and cluster thinning also may be needed on vines with heavy loads of fruit.

Training Systems

The grape vine is quite adaptable, and the grower has a choice of many different training systems. The more popular systems for 'Concord' and other American cultivars are the four-cane Kniffin, the umbrella Kniffin, and the high cordon. The high cordon system is preferred for most French hybrids because of the upright-growth characteristics of their new shoots.

Training System: Four-Cane Kniffin

At planting time, prune the vines to a single stem with two buds (Fig. 24, page 52). Shoots should grow from each bud. If the trellis is not yet constructed, tie the most vigorous shoot to a stake four to five feet tall. The tip of the second shoot can be pinched back. At the end of the first summer, the main shoot should be three to four or more

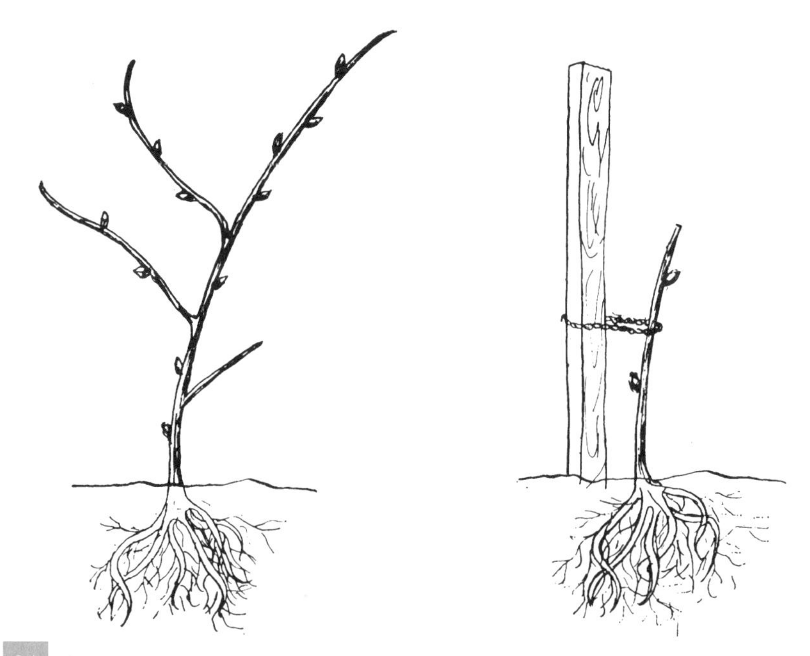

24 A newly planted grapevine is shown before (left) and after (right) pruning.

feet high, and may be long enough to reach the top wire of the trellis.

Four-Cane Kniffin Training System: Second Year

In the early spring while the vine is still dormant, prune off all but the strongest cane. Tie the cane tautly to the top wire of the trellis or to the lower wire if it is not long enough to reach the top wire. This cane will become the permanent trunk.

During the second growing season, remove shoots that develop below the lower wire and remove any flower clusters. Allow the main trunk to reach the top trellis wire, and allow some short lateral canes to develop along each wire (Fig. 25).

Third Year

If one to four strong lateral canes develop during the second year, they may be trained to trellis wires. Otherwise, cut the vine back to a single vertical trunk. If short canes are present, leave two buds on each of two shoots near the lower and upper trellis wires. Fruiting canes for the next season grow from these buds. If the vines are strong, a small crop can be grown. If the vines are still small, remove any flowers.

During the third summer, numerous lateral canes develop that should bear a good crop during the fourth year.

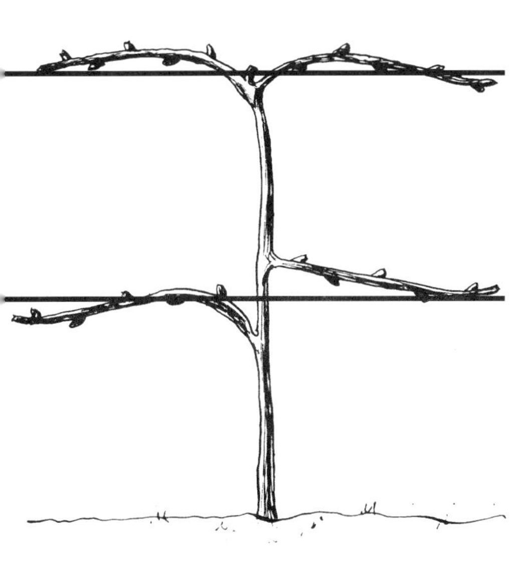

Mature Vines

After the third year, the vines can be treated as mature vines (Fig. 26). In early spring, prune the vine to four one-year-old lateral canes, called arms. Each arm will have six to twelve buds for a total of thirty-five or more buds on a vigorous vine. Each of these buds is capable of producing two or three clusters of grapes. Leave two renewal spurs near the main trunk for future fruiting canes at each trellis wire. Remove all other growth.

Select canes of moderate vigor for the lateral fruiting canes. They should be one-third to one-half inch in diameter, straight, and preferably unbranched. Do not select canes less than one-fourth inch in diameter, or "bull canes" that are long, heavy, more than one-half inch in diameter, vigorous,

25 (above) A grapevine after two growing seasons. On vigorous plants, four lateral canes may develop, and four to six buds may be left on each lateral cane.

26 (below) A grapevine after three growing seasons. A maximum of twelve to fifteen buds may be left on each lateral cane.

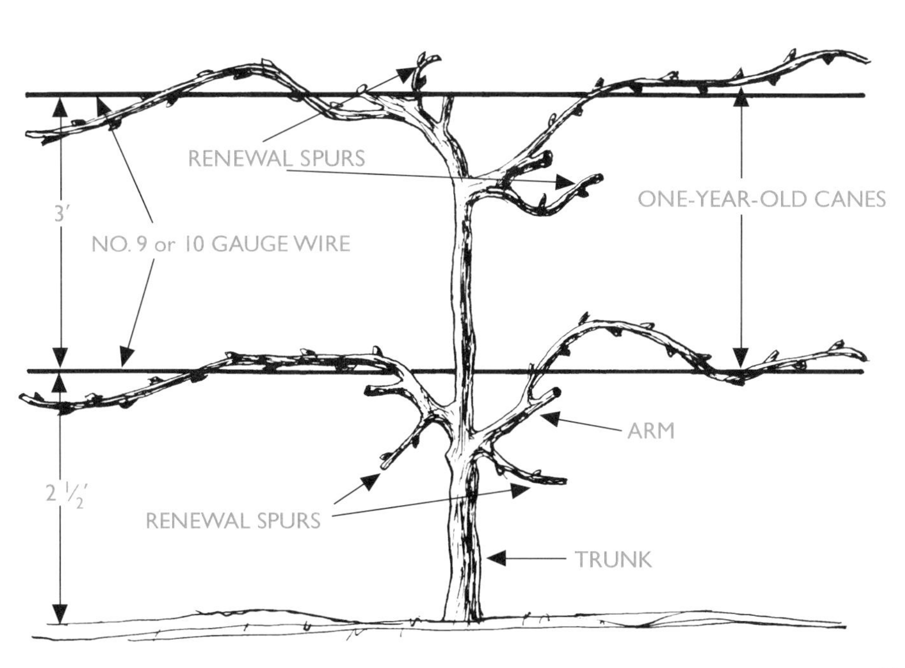

and tend to produce little fruit. Train one cane each way on the trellis wires. These lateral canes should originate from the main trunk or as near to it as possible on the trellis to form the "four arms."

After pruning, loop or spiral the canes over the support wires and tie with twine or other durable material (Fig. 26, previous page).

Training System: Umbrella Kniffin

This system is a modification of the four-cane system and can be used on two-wire or three-wire trellises. The trunk is terminated six to twelve inches below the top wire. Three to six canes are shortened to six to ten buds each, are bent over the top wire, and are tied to a lower wire. Two or three other canes are shortened to two buds to serve as renewal spurs and to provide the fruiting canes for the following year's crop. All other canes should be removed. The umbrella Kniffin system can be used for both American and French hybrid cultivars (Fig. 27).

27 Umbrella Kniffin training system. Prune long, with six to ten buds on each cane.

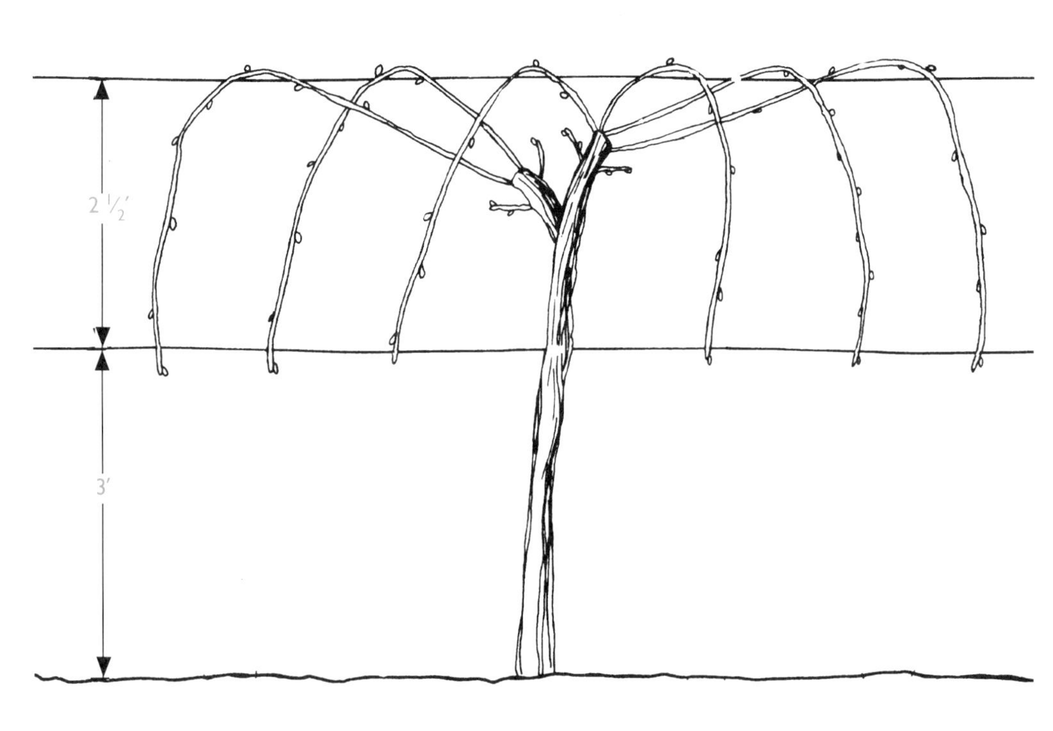

Training System: High Cordon

Ease of pruning is the outstanding characteristic of the high cordon system. In this system, the trunk or trunks are trained to a single wire five to six feet from the ground to form cordons or semi-permanent arms (Fig. 28). The trunk is trained to the wire and two arms are set permanently to form a T-shaped vine. Canes growing from the arms are selected and are "spur pruned" to three to four buds each. The vine will have a total of forty to sixty buds. Remove excess canes and spurs. During the summer, shoots arising from the fruiting canes are "combed" or positioned to grow downward to form a curtain. Canes growing vertically above the cordon can be removed.

American types with droopy growth habits are well adapted to the high cordon system. French hybrids also can be grown with this system.

28 **High cordon training system.** Prune shoots to form many two- to four-bud spurs on each arm.

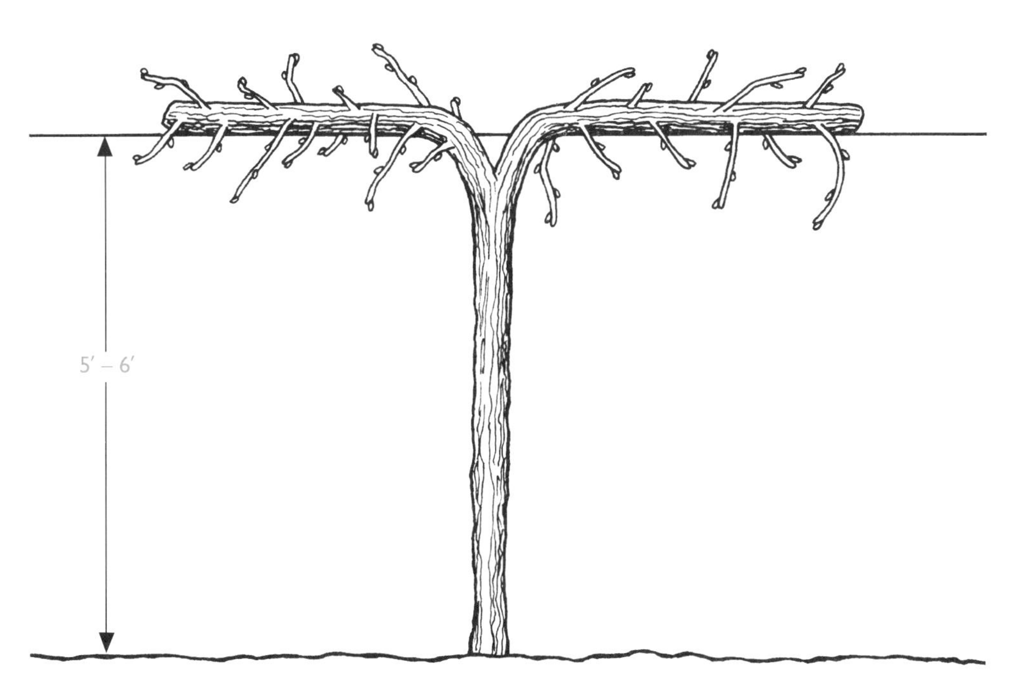

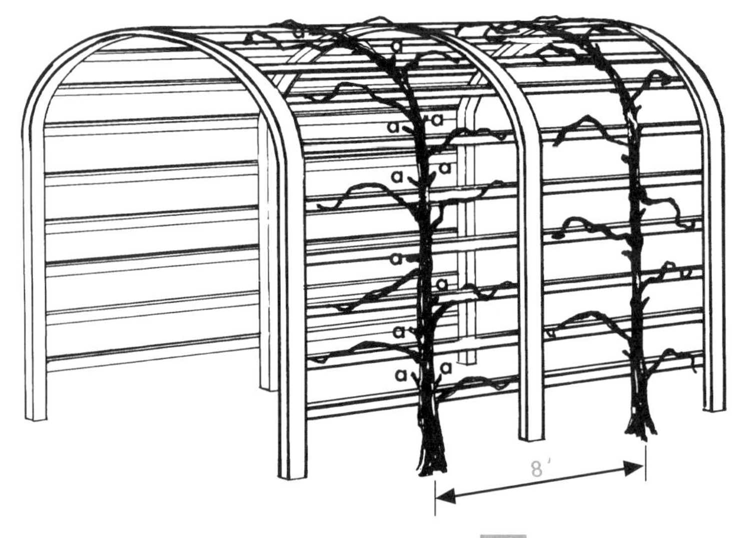

29 Mature grapevines trained and pruned on an arbor. Note renewal spurs.

Training and Pruning on Arbors

Grapes make excellent dual-purpose vines when trained on arbors, on a pergola for a summer roof, on a fence or wall—almost anything will work. Good varieties, properly cared for, will produce good-quality fruit as well as shade or screen for the home landscape.

An endless variety of sizes and designs is used as supports for grapes. Whatever the design, use materials that permit long-time use with a minimum of repair. Wood, metal, masonry, or a combination of these is appropriate. Wire mesh is not recommended: it makes a good support for the vines, but training and pruning operations are extremely difficult. Allow at least eight feet between vines. Spacing too close results in a junglelike growth that favors the development of diseases, interferes with satisfactory fruiting, and makes pruning difficult. When designing the structure and subsequently training and pruning, keep in mind that the reason for growing grapes this way is to combine fruit production with shade or screen.

If overhead cover is desired on arbors or pergolas, the permanent single trunk is trained to the top of the structure (Fig. 29). Each year, one-year-old canes three or four feet long are distributed at intervals of three feet along this permanent trunk. Renewal spurs of two or three buds are distributed along the trunk in the same manner to produce new fruiting wood from which to select the one-year-old canes for the following year. This type of pruning is not severe enough for best fruit production, but it results in the desired shade. As with other systems, the amount of bearing wood needs to be adjusted for the vigor of the vine, but remember that most home arbors soon become a tangled mass of vines because of failure to prune heavily enough.

30 2,4-D damaged grape leaves.

Pruning Neglected Vines

Old vines that have not been pruned for two or more years will be dense and may have several stems or trunks arising from the roots. It may take two or more years to salvage neglected vines. Basically, the job is to select a frame (main trunk, fruiting arms, and renewal spurs) as described earlier. If neglected vines have multiple trunks, remove several each year until one to three remain. Select young canes from the base for the future trunks if the older trunks are malformed. Desirable fruiting wood may be far from the main trunk. Prune out the older portion of the plant gradually, while maintaining moderate fruit production, until the new trunks are ready to take over.

Chemical Injury

Grapes are one of the most sensitive plants to many chemicals, particularly herbicides containing 2,4-D. *Do not apply* fertilizer or herbicide that contains 2,4-D to the lawn near grapes. Avoid using any 2,4-D (or sprayers that have contained 2,4-D) in the vicinity of grapes. If these precautions are not taken, enough 2,4-D can drift one-half to one mile by air and can ruin your grape crop. Injury may be indicated by misshapen leaves, tendrils, and young shoots. The leaves may have sawtooth edges and may be narrow and fan shaped. The clusters may ripen unevenly or not at all. The symptoms appear one to three weeks after exposure to the fumes (Fig. 30).

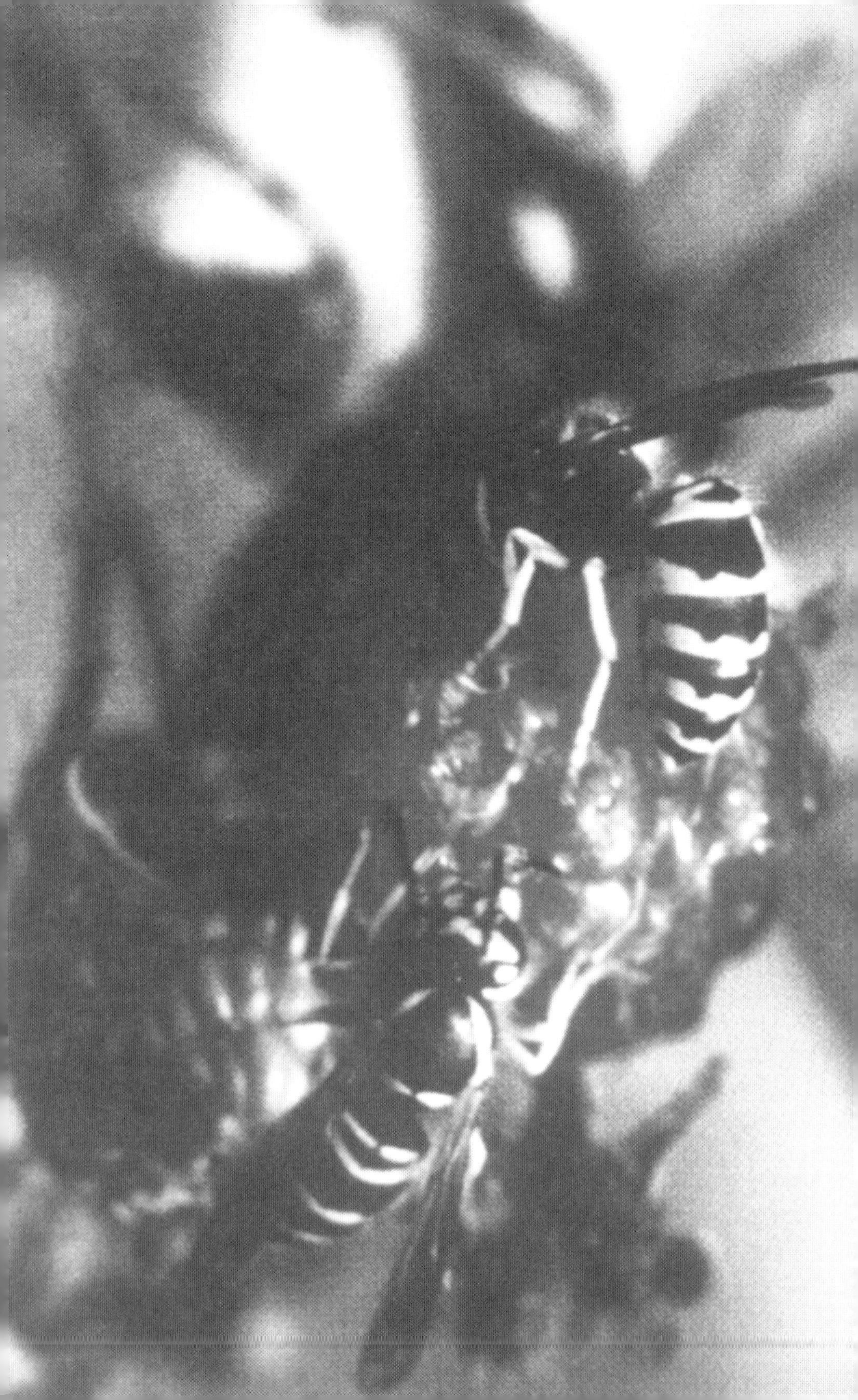

Pest Control

To have a successful fruit garden, you must be aware of many diseases and insect pests and be prepared to control them. Home fruit growing is usually an avocation rather than a vocation, and a person often does not have the time or inclination to spray pesticides. Thus, most pest-control programs for the home fruit garden are quite limited, although they can also be extremely successful.

Pest control begins with the selection of a suitable planting site and consideration of previous cropping history, followed by selection of disease-resistant cultivars, use of healthy disease-free planting stock, and, finally, the use of good cultural and sanitation practices. These factors are discussed on the following pages as they apply to each crop. The use of pesticides to control certain insects and diseases is often necessary because sanitation practices are not always adequate by themselves. One well-timed chemical application may mean the difference between a good crop and no crop at all. (*Home Fruit Pest Control* and *Illinois Homeowner's Guide to Pest Management* contain spray recommendations. See page 77 for further information about these publications.)

(left) Yellowjackets attacking ripe red raspberry fruit.

Strawberries

Strawberries can be extremely successful when a small number of precautions are taken against pests. The most damaging insects are the strawberry weevil, which is called "clipper" because it cuts off the buds and fruit, leaving them hanging as though they had been partly broken off; the "cat-facing" insects (many different sucking insects) that produce a deformed, low-quality berry; the crown borer, a white, thick-bodied grub, one-fifth inch long, that feeds inside the plant at the soil level so that the strawberries die or are so weakened as to be susceptible to other pests; and white grub, the large grub up to one inch long that kills the plant by feeding on the roots. The important diseases are gray mold and leather rot, which kill the blossoms and rot the fruit. Anthracnose, red stele, and Verticillium wilt diseases are caused by soil-borne fungi that infect the roots and kill the plant. Leaf spot, leaf scorch, and leaf blight are three important foliar diseases. The recommended chemicals and spray schedules are given in *Home Fruit Pest Control* (see page 77). Many of these diseases can be avoided by using resistant cultivars.

Strawberries: Sanitation

Use of cultural and sanitary practices is of utmost importance in controlling various diseases and insects affecting the strawberry. The following practices are recommended:

1. Begin by planting healthy plants on fertile soil that has been plowed deep and contains 3 to 5 percent organic matter.

2. Obtain virus-free plants for new plantings. Reliable nurseries have virus-free plants readily available. For white grub control, plant strawberries on land that has been under cultivation for one or two years, if possible.

3. If crown borer has been a problem, locate new beds as far as possible (more than three hundred feet) from old beds, and plow infested land immediately after harvest.

4. Choose disease-resistant cultivars.

5. Renovate beds immediately after each harvest (see page18).

6. Since Verticillium wilt is a soil-borne disease that enters through the roots, avoid planting strawberries after other crops that are susceptible to this disease, such as tomatoes, potatoes, eggplant, melons, okra, beets, peas, brambles, and roses.

7. Control nearby weeds that may harbor insects and certain pathogens.

8. Do not over-water, under-water, or over-fertilize plants.

9. Provide good air circulation.

Brambles

Brambles: Sanitation

Many diseases of brambles cannot be controlled by pesticides. These include crown and cane gall (bacterial diseases that produce large tumorlike growths on the roots and canes) and five viral diseases that are characterized by their names—mosaic, leaf curl, ring spot, bushy dwarf, and streak (dark-blue or violet-blue markings that appear longitudinally on the canes from the ground up).

The following sanitary procedures must be observed:

1. Wait at least three years to replant a site where crown and cane gall or Verticillium wilt-infected plants have grown. To protect against Verticillium wilt (caused by a soil-borne fungus that infects the roots), do not plant brambles where tomatoes, potatoes, eggplant, melons, okra, beets, peas, roses, or strawberries have grown within the past three years.

2. Select virus- and aphid-resistant cultivars.

3. Order plants from a reliable nursery.

4. Purchase certified virus-free planting stock.

5. At planting time, cut off old stubs of nursery stock and "handles" of young purple and black raspberries. These "handles" are left on the plant by the nursery to facilitate handling. The old stubs may be infected with anthracnose, a serious fungal disease of many brambles.

6. As soon as plants with a disease that cannot be controlled by chemicals are detected, dig the plants with as many of their roots as possible, and destroy them by burning or placing them in a trash can.

7. Prune out canes that do not leaf-out normally in the spring.

8. Immediately after the summer harvest, remove and burn or compost all canes that have fruited. Do not return this compost to the planting. Also, cut out surplus or weak canes and those showing injury. Canes of everbearing red or yellow cultivars grown under the one-crop (primocane-bearing) system are not removed after the fall harvest.

9. Remove nearby (within three hundred feet, if possible) wild brambles and neglected plantings; they are a source of disease and insect pests.

10. Tie long canes, especially semi-erect blackberries, to supports. This practice allows better air movement and sunlight penetration, which helps in disease control.

11. Keep plantings and surrounding areas free of weeds.

Brambles: Spray Schedule

Use the spray schedule for brambles (raspberries and blackberries) outlined in *Illinois Homeowner's Guide to Pest Management* or *Home Fruit Pest Control* (see page 77).

Blueberries

Because blueberries are a relatively new crop in Illinois in terms of acreage, the pests affecting them have not assumed major importance. If pest control is needed, follow the spray schedule for blueberries in *Illinois Homeowner's Guide to Pest Management* or *Home Fruit Pest Control* (see page 77). Some of the potential problems that have been observed in Illinois are mummy berry, Phomopsis cane blight, fruit worms, and scale.

The best way to care for blueberry plants is to prune them annually, use mulch, irrigate properly, and give them an early dormant oil spray to control scale and mites.

Currants & Gooseberries

These fruits require a minimal spray program. The important insects are the currant aphid, which causes bright red, cupped, distorted, or wrinkled areas on the leaves; the imported currant worm, about one inch long and greenish with black spots, which feeds on the edges of the leaves; the cane borer, a worm one-half inch long that burrows the entire length of the cane and dwarfs the plant; and scale insects whose very small, grayish, nipple-shaped coverings can be seen on the bark. The important diseases include:

- anthracnose and leaf spot, which cause a spotting of the leaves and a yellowing most pronounced on the gooseberry,
- cane blight, which may suddenly wilt and kill scattered canes or bushes (cut out such canes and destroy them immediately), and
- powdery mildew, particularly on gooseberry, which forms white patches on the surface of leaves, shoots, and berries, eventually distorting them.

For these and other insect and disease problems, the spray schedule outlined in *Illinois Homeowner's Guide to Pest Management* or *Home Fruit Pest Control* (see page 77) is effective.

When pruning currants and gooseberries during the dormant season, watch for signs of cane borers (hollow canes with black centers). Remove such canes and destroy all prunings immediately. Keep the plants from becoming too dense (see page 45 for pruning instructions). This practice will reduce damage from diseases by allowing air circulation and quick drying of the foliage.

Grapes

Because grapes have many diseases and most cultivars have little resistance, grapes must be sprayed. Spraying usually provides excellent control of troublesome pests that occur in most seasons. The grape berry moth, which causes wormy fruits, and leafhoppers, which are small brightly colored insects that feed on the leaves and produce tiny whitish spots and sickly appearing plants, are the most troublesome insects. A serious insect problem of some American and many French hybrids is the grape phylloxera. The important fungal diseases are black rot, which causes brownish circular leaf spots and black, wrinkled, dried-up fruit; and powdery and downy mildews, which result primarily in a scorched appearance of the leaves. For spray schedules, see *Illinois Homeowner's Guide to Pest Management* or *Home Fruit Pest Control* (see page 77).

For Your Protection

Always handle pesticides with respect. After all, the people most likely to suffer ill effects from pesticides are the applicator and his or her family. Accidents and careless, needless overexposure can be avoided. Every year, there are deaths due to accidental ingestion of pesticides.

Each year, more than 750 Illinois children under 12 years of age are rushed to a doctor because of suspected pesticide ingestion or excessive exposure. A study of such cases showed that 50 percent of the children obtained the pesticide while it was in use and 13 percent obtained it from storage. Fifty-three percent of the cases involved pesticides used as baits. These accidents could have been prevented. The following suggestions for safe use of pesticides are designed to prevent such unfortunate, careless accidents.

1. Store pesticides out of reach of children, irresponsible persons, or animals; store in a locked cabinet away from food or feed.
2. Put pesticide containers back in the storage area before applying the pesticide. Children have found open bottles by the water tap.
3. Avoid breathing pesticide sprays and dusts over an extended period. This is particularly true in enclosed areas such as crawl spaces, closets, basements, and attics.
4. Wash with soap and water exposed parts of the body and clothes contaminated with pesticide.
5. Wear rubber gloves when handling pesticide concentrates.
6. Do not smoke, eat, or drink while handling or using pesticides.
7. Do not blow out clogged nozzles with your mouth.
8. Leave unused pesticides in their original containers with the labels on them and in locked cabinets.
9. Wash out empty pesticide containers three times and then bury them or place them in the garbage.
10. Do not leave puddles of spray on impervious surfaces.
11. Do not apply pesticides to fish ponds, bird baths, or pet dishes.
12. Do not apply pesticides to dug wells, cisterns, or other water sources.
13. Observe all precautions listed on the label. Use pesticides only on the *crops* specified, in *amounts* specified, and at *times* specified.

Edible Landscapes

Landscaping a new home or renovating an existing landscape performs several functions. Landscaping beautifies the home, increases the value of the property, and helps the area surrounding the home to function as an extension of the indoor living space. Traditionally, these objectives are met by using various types of ornamental plants—ground-covers, both deciduous and evergreen shrubs, small flowering trees, and shade trees. But...

...those who dare to be different from the average homeowner can incorporate plants producing edible products into the residential landscape. Many of the functions of landscape plantings—screening, defining space, providing interest, and aesthetics—can be suitably served by fruit-producing plants. These plants can provide the homeowner with the enjoyment and relaxation of gardening and with the satisfaction of producing fresh, high-quality fruit for the entire family at low cost.

It must be emphasized, however, that fruit plants require higher maintenance than do the more traditional landscape plants. To produce a satisfactory crop, fruit plants must be regularly fertilized, mulched, watered, pruned and trained, sprayed for pests, kept weeded, and harvested. Without these important operations, only low-quality produce can be expected.

Designing with edible, fruit-producing plants should be done on a gradual basis. Add a few plants annually until your limit of upkeep is reached. Limiting fruit plants to the outdoor living area—the back or side yards—may be a wiser choice than using them on the entire property. This will avoid creating a landscape whose maintenance exceeds the capabilities of the homeowner.

This grape plant (left) both produces fruit and provides shade for the patio.

Food-producing plants can be gradually integrated into an established landscape with traditional plantings. Landscaping for new properties can be completely designed but installed in stages, amending plant choices as maintenance limits become apparent. In all cases, the fruit plants should be chosen to perform the same landscape functions as the traditional plant materials. If a large shrub is required, you should not plant an apple tree just to have the fruit.

The most important step in landscaping with fruit plants is proper plant selection. Suitability to the site and performance in the particular climate are the main considerations. Choose the proper size and shape for the location, and then determine which fruit plants satisfy these criteria. This being done, a cultivar can be selected. Fruit plant cultivars differ in their temperature hardiness, soil tolerance, moisture requirements, and disease resistance. Careful planning before plants are purchased will reduce future maintenance problems and total cost.

Groundcovers

Grasses are the most widely used groundcovers. They are easily and rapidly established, can cover wide expanses, are fairly easy to maintain, and can withstand foot traffic. Other kinds of plants are used on steep banks, under trees, and in shrub beds. These include selections such as ivies, vinca, and low-growing evergreen and deciduous shrubs. Runner-producing plants like strawberries and low-growing vines such as dewberries can replace traditional groundcovers. Dwarf blueberries can be established in acid soil. These plants require annual or biennial mowing or pruning.

Container or Patio Plants

Plantings in containers or tubs are often used for decks, patios, and around homes on very small lots where bed space is limited. Commonly used pot plants are annual bedding plant types such as geraniums, petunias, marigolds, and begonias. Pots of dwarf blueberries, figs, strawberries, or even grapes can be used as replacements for these small, decorative container plantings.

Small trees and shrubs are often used in large containers or tubs when a greener, more lush effect is desired. These can be replaced by dwarf fruit trees such as the 'North Star' tart cherry or fully dwarfed, spur-type apples. Even though small in stature, these trees will produce high-quality, standard-sized fruit.

Flower Beds and Borders

Flowering plants are difficult to replace with fruit crops. The flowers of fruit plants are beautiful individually and when viewed close up, but their floral display cannot rival that of the tried-and-true annual and perennial bloomers. Flower colors of fruit plants are generally limited, with most being white to pink and only a few ranging into the darker reds. They also typically bloom only once a year and then only for a short period of time. Because of the narrow range of colors, the small

flower size, and the seasonal bloom of most fruit plants, the homeowner who likes the "showy" display of flowers in beds and borders should continue to rely on the wide variety of annuals and perennials available for this purpose. Remember, however, that strawberry plants make a neat and attractive dual-purpose border for ornamental beds.

Hedges and Shrubs

Evergreen and deciduous shrubs, which are by far the largest group of landscape plants, come in a multitude of sizes, shapes, and colors. Such shrubs can be used in almost any landscape situation as screens, windbreaks, shrub borders, hedges, group plantings, "living walls," or individual specimens. Included in this group are the honeysuckles, the viburnums, euonymus, juniper, yew, and many more.

A large variety of fruit crops is suitable as replacements, satisfies the landscape requirements, and produces sizable yields of fresh fruit. Currants, gooseberries, blackberries, filberts, and quince make interesting additions to shrub borders. They can also be used as single specimens or in masses. Be sure, however, that their looser, more spreading form is acceptable for the landscape situation.

Highbush blueberries require a cool, moist, acid soil and consequently do well when planted in conjunction with acid-loving rhododendrons, azaleas, and andromeda. Hedges of red raspberries and thorned blackberries are easy to establish due to their free-suckering roots. The serviceberry, a common ornamental, also produces delicious blue-black fruit in June.

Trees

Trees can be divided into two subgroups: small ornamental trees (eight to twenty feet tall), and large shade trees (twenty feet and over). Flowering dogwood, eastern redbud, amur maple, star magnolia, and the popular crabapples are small trees that are frequently used as single "showy" specimens, or placed near patios and decks for their beauty and fragrance. The best use of this group is in corner plantings to integrate the bold lines of architecture into the landscape beyond.

Dwarf and semi-dwarf fruit trees—apples, pears, peaches, cherries, and plums—can readily replace the common small tree choices mentioned above. The mature size can be adjusted to the site by careful selection of scion/rootstock combinations. Certain large-fruited varieties of crabapples are good for making jellies and preserves. The fruit of the pawpaw is a special treat, too.

For trees in the large shade-tree category, tall-growing nut trees such as persimmon, walnut, pecan, butternut, blight-resistant chestnuts, and hickory can be planted. However, the decision to plant large fruit trees in the suburban landscape must be carefully thought through. Large trees are difficult (and dangerous) to harvest and can pose a real maintenance problem for the average homeowner.

Cooking & Eating the Fruits of Your Labor

After having set your plants in the soil and having tenderly cared for them for several years to produce enough fruit for you to use, you are now ready to harvest and eat the berries! But when are the fruit ready to pick?

Are They Ripe?

The best ways to determine when the fruit is ripe enough are by looking at them and tasting them. As the crop ripens, look over the fruit at frequent intervals. Berries signal their readiness to be harvested and cooked into delicious family meals by changing their colors. Strawberries and raspberrries change from white to red, blackberries from red to black, and grapes from green to blue, red, or yellow-green.

When they look ripe and have the right color, pick several berries from the plants and taste them. If the fruit is soft enough and sweet enough for your taste, harvest the ripe berries. Once the berries start to ripen, harvest every two to three days. Strawberries and raspberries can become over-ripe and start to rot within a few days.

Time to Harvest

Harvest in the morning, if possible. At this time, the berries are still cool from the night and will store better. Picking should be done gently; never squeeze the berries. Pick strawberries, currants, and grapes by the stems. Pick the bramble fruits by placing your finger-tips around the base of the berry and gently pulling the berry from the plant. Use your thumb to "roll" blueberries from the cluster into your palm.

Containing and Storing Berries

Place all the ripe berries in containers that are no more than three to four inches deep. Large containers such as buckets are not recommended; the fruit in the bottom of deep containers will be squashed and leak juice. Set picked berries in the shade if large quantities are being harvested. As soon as possible, store the berries in a refrigerator to remove the heat from them. If the berries are to be eaten fresh, wash and drain them before enjoying the fresh flavors.

Freezer Jams Preserve Flavor

Freezer jams made according to pectin manufacturers' (such as Surejell) recipes are an excellent way to preserve the fresh flavors of soft fruits. Small fruits and grapes can also be frozen easily, but do not wash them prior to freezing. Washed berries tend to become very mushy upon freezing.

Blueberries, raspberries, and blackberries need only be examined for unwanted stems, over-ripe fruit, or other materials before placing in a bag for freezing. Strawberries, currants, and gooseberries should have the stems removed before freezing. Grapes should be removed from the stems and frozen in a light sugar syrup for best results.

Recipes for Small Fruits

1/4 c. softened butter or margarine, or
1/4 c. oil for those watching cholesterol

1-3/4 c. flour

2-1/2 tsp. baking powder

1/2 tsp. salt

1/2 c. sugar

1 large egg or egg substitute

1/2 c. milk

1 c. washed and drained blueberries

Most Delicious Blueberry Muffins

Mix dry ingredients together. Add softened butter, then unbeaten egg, and finally the milk. Stir until batter is smooth. Gently fold in the drained blueberries. Spoon batter into greased muffin pan. Bake at 375° F for 20 to 25 minutes. Remove muffins from the pan, split while still hot, and place a pat of butter inside to melt. Great for Saturday morning breakfast.

Makes 12 large muffins.

Stawberries or Raspberries

There are two best ways to eat these soft fruits. The first way is fresh, on a good homemade biscuit shortcake. The second is as frozen homemade jam—in the winter on top of thick slices of homemade bread or toasted English muffins.

Grandma's Shortcake for Strawberries or Raspberries

Clean at least 3 pints of fruit. Remove stems of strawberries and slice or chop them into small pieces. With raspberries, simply wash and drain. Place berries in a bowl and add 1/2 to 1 cup of sugar, depending on the sweetness of the fruit and individual taste. Stir the sugar and fruit together and place in a refrigerator to cool. The berries should become juicy.

Then put together the shortcake:

2 c. flour

1 tsp. salt

4 tsp. baking powder

2 Tbsp. sugar

5 Tbsp. butter or shortening

2/3 c. milk

Mix dry ingredients together. Work in the butter with a pastry blender and then stir in the milk. The dough will be stiff. Place in a greased pie pan and bake at 400° F for 15 to 20 minutes.

Remove from the oven. Spread butter on the hot surface and allow the shortcake to cool for 15 or 20 minutes. During that time, whip one cup of cream. Split the cooled shortcake into two layers. Place half of the cool, juicy berries from the refrigerator on the bottom layer, cover with the top layer, and place the remaining berries on top. Serve the whipped cream on individual servings.

This dish is good served for lunch, omitting the whipped cream and accompanying the shortcake with slices of a hard cheese, such as Jarlsberg or a mild Swiss and hot or iced coffee.

Currants

Red and white currants are seldom used in cooking. Do not confuse these with the dried currants found in grocery stores. Dried currants are actually small raisins and not the same as the currants discussed here.

Apple-Currant Pie

This is a simple variation of the traditional apple pie that most people enjoy.

Begin as you would when making a typical 9-in. or 10-in. deep-dish apple pie. Make the crust, peel and slice 6 to 8 large apples (preferably a flavorful sweet-tart apple such as a 'Jonathan'), and place the apples in the pie crust.

Then, instead of adding spices to the pie, add 1 cup of fresh or frozen red or white currants on top of the apples. Sprinkle approximately 2 tablespoons of cornstarch, 1 cup of sugar, and 1/4 teaspoon of salt over the fruit. Cover with the top crust and bake as usual. The currants will add a spicy flavor to the apple pie.

Orange-Currant Chicken

Another use for red currants is to make delicious red currant jelly. This jelly can be eaten with breads or meats and can be used in pastries. This recipe blends spiciness and sweetness to change the taste of traditional fried chicken.

One whole chicken, cut in pieces and skinned, or three whole breasts

1 c. orange juice

2 Tbsp. currant jelly

2 Tbsp. prepared ground horseradish

Brown the chicken pieces in a hot frying pan in a small amount of oil. Reduce the temperature and add the other three ingredients that have been combined in a blender. Cover and let the chicken simmer for 30 minutes. Turn the meat occasionally for the flavors to penetrate all of the meat.

Grapes

What can be done with grapes except for making jelly, juice, and an occasional grape pie? One old-time recipe is grape pudding. This pudding is similar in texture to chocolate or butterscotch puddings, but tastes like grapes.

Grape Pudding

Select 2 quarts of grapes (the best are a blue grape with a full flavor such as 'Buffalo,' 'Concord,' or 'Fredonia.' Other cultivars can be used to change the color and flavor intensity).

Cook the grapes with 1 cup of water and extract the juice. There should be about one quart of juice. Add 1/2 cup of sugar to the juice — more if the juice is not sweet enough.

Heat the juice and add 3 tablespoons of cornstarch or tapioca. Dissolve the cornstarch in a small amount of water before stirring into the juice. Cook until thickened and pour into a serving bowl. Top with sweetened whipped cream.

Blackberries

Blackberry pie is one of the best pies to eat. Seedless blackberry jelly is also very good but takes a lot of time to prepare. To enjoy blackberries easily and quickly, make an old-fashioned cobbler using biscuit dough.

Blackberry Cobbler

Wash 1 quart of blackberries and drain well. Mix the berries with 2 tablespoons of cornstarch and place in the bottom of a baking dish. Pour 3/4 cup of sugar on top of the berries with a pinch of salt. Then, prepare biscuit dough the same as shortcake (page 72) with these exceptions: increase the sugar to 3 tablespoons and the milk to 3/4 cup. Drop or spread dough on top of the blackberries and sprinkle extra sugar on top. Bake for 25 to 30 minutes at 350° F or until done.

Gooseberries

These berries are enjoyed by Europeans much more than by Americans. The berries can be left on the plants until very sweet, but most recipes call for fruit that is not fully ripe and somewhat tart. The following recipe is a modification of the trifles that are common in England.

Gooseberry Trifle

1 qt. gooseberries, stemmed and washed

1-1/2 c. water

3/4 c. sugar—more or less according to the tartness of the fruit

Cook the berries in the water slowly until they are soft. Drain the fruit, but reserve the cooking liquid. Puree the berries in a blender to remove the seeds. Measure the pureed fruit and add enough of the reserved liquid to make 2-1/2 cups. Add the sugar and stir until sugar is dissolved. Cool in the refrigerator.

Next make a custard with:

1 c. milk

3 egg yolks or 1/4 c. of egg substitute

1 tsp. cornstarch

Heat the milk slowly. Combine the cornstarch with the egg yolks in a small bowl and whisk into the warm milk. Continue to cook while stirring until the custard is thick. Simmer about 2 minutes more. Cool slightly and stir into the gooseberry puree. Chill this mixture.

Whip 1 to 1-1/2 cups of whipping cream sweetened with 3 tablespoons of powdered sugar to stiff peaks and fold gently into the gooseberry-custard mixture.

Line a 2-quart bowl with thin slices of sponge cake, pound cake, or lady fingers. Sprinkle with orange juice or orange flower water. Pour the gooseberry mixture over the cake slices, top with lightly toasted cake crumbs, and serve.

Fruit Soup

When you have leftover fruits, either fresh or frozen, but not enough to make a particular dish, do as the Scandinavians and mix the fruits together into a sweet soup. This soup is eaten hot or cold, at the beginning of a meal or at the end as a dessert with cookies. Fruits can be combined in any proportion. However, any of the small strawberries, when cooked, will have a very mushy consistency. This dish is better if they are not included.

1 qt. mixed berries (except strawberries)

1/2 c. sugar, or to taste

1/4 lemon sliced very thin

1 cinnamon stick

1 Tbsp. cornstarch or tapioca

Heat the berries with about 1/2 cup of water over a medium heat. Stir in the sugar. Add the lemon and cinnamon stick and cook for 15 minutes. Mix the cornstarch with a small amount of water and blend the paste into the fruit mix. Cook until the fruit is thickened. If too thick, the soup may be thinned with water or fruit juice. Serve either hot or cold.

The following references are listed for gardeners seeking more detailed information on cultivars, culture, propagation, and disease and insect control related to the small fruits.

BOOKS

Modern Fruit Science (1995). N.F. Childers, J.R. Morris, and G.S. Sibbert. Horticultural Publications, Gainesville, FL.
Small Fruit Crop Management (1990). G.J. Galletta and D.G. Himelrick. Prentice-Hall, Englewood Cliffs, NJ.
Temperate Zone Pomology (1993, 3rd ed.). M.N. Westwood. Timber Press, Portland, OR.

CIRCULARS AND BULLETINS

The publications listed below may be purchased from Information Technology and Communication Services, College of ACES, University of Illinois at Urbana-Champaign, 1401 S. Maryland Drive, Urbana, IL 61801 (toll free: 800-345-6087).

C1354 *Illinois Homeowner's Guide to Pest Management* (1997, $8)
IDEA2 *Small Fruits: Insect and Disease Management for Backyard Fruit Growers in the Midwest* (1997, $5)

The publication listed below may be obtained by contacting J.D. Kindhart, University of Illinois, Dixon Springs Agricultural Center, Route 1, Box 256, Simpson, IL 62985.

C1262 *Illinois Commercial Small Fruit and Grape Spray Guide* (issued annually)

The following publication is available on the VISTA Web Site.

C1145 *Home Fruit Pest Control* (1993)
(http://www.ag.uiuc.edu/~vista/abstracts/aHOMEFRUT.html)

The following publication is available at The Ohio State University Extension Bulletin Web Site. Soft- and hardbound copies can be purchased from Extension Publications at The Ohio State University, (614)292-1607.

BULLETIN 861 *Midwest Small Fruit Pest Management Handbook*
(http://www.ag.ohio-state.edu/~ohioline/b861/)

PROCEEDINGS

Proceedings of the Illinois Small Fruit and Strawberry School are issued annually. Many back issues are available. Contact J.D. Kindhart, University of Illinois, Dixon Springs Agricultural Center, Route 1, Box 256, Simpson, IL 62985.

NEWSLETTERS

To subscribe to *Fruits and Vegetables Newsletter*, contact R. Weinzierl, Department of Crop Sciences-Entomology, University of Illinois at Urbana-Champaign, AW-101 Turner Hall, 1102 S. Goodwin Avenue, Urbana, IL 61801. E-mail subscriptions are available. 25 issues per year.

OTHER UNIVERSITY OF ILLINOIS PUBLICATIONS

The publications listed below may be obtained from the Department of Natural Resources and Environmental Sciences, NRES/Horticulture, University of Illinois at Urbana-Champaign, W-503 Turner Hall, 1102 S. Goodwin Avenue, Urbana, IL 61801.

FR-1-79 Propagating Grapes from Cuttings
FR-2-80 Sources of Small Fruit Plants
FR-3-80 Distinguishing Thornless Blackberry Cultivars
FR-4-84 The Fall-Crop-Only Cultural System for Everbearing Red Raspberries
FL-13-81 Taking Soil Samples in the Yard and Garden

The publications listed below may be obtained from the Department of Crop Sciences/Extension Crop Pathology, University of Illinois at Urbana-Champaign, N-533 Turner Hall, 1102 S. Goodwin Avenue, Urbana, IL 61801.

FL 1 Strawberry Spray Schedule
RPD 700 Raspberry Anthracnose
RPD 701 Strawberry Red Stele Root Rot
RPD 702 Strawberry Leaf Diseases
RPD 703 Black Rot of Grape
RPD 704 Gray Mold of Strawberries
RPD 705 Downy Mildew of Grape
RPD 706 Leaf Variegation in Strawberries
RPD 707 Verticillium Wilt of Strawberry
RPD 708 Orange Rust of Brambles
RPD 709 Spur Blight and Cane Blight of Raspberries
RPD 710 Virus Diseases of Brambles
RPD 711 Failure of Fruit Set in Blackberries
RPD 1006 Crown Gall
RPD 1010 Verticillium Wilt Disease